MW01233572

MakerBot.com

ISBN: 978-0-9991345-0-4

Printed in the United States of America

First Edition

10 9 8 7 6 5 4 3 2 1

MAKERBOT EDUCATORS GUIDEBOOK

The Definitive Guide to 3D Printing in the Classroom

WHY WE WROTE THIS BOOK

In 2008, while we were still building our first 3D printer, MakerBot founded Thingiverse® —a 3D file library and community. Over the years, as we engineered more and more printers, we watched Thingiverse grow into a massive, indispensable resource for teachers, designers, and makers. From that moment on, educators were hungry for guidance on how to introduce 3D printing into their classrooms. So in 2014 we wrote the industry's first how-to, MakerBot in the Classroom.

By then, the 3D printing movement in education had exploded and showed no signs of slowing; by 2016 MakerBot® 3D printers could be found in over 5,000 schools worldwide.

After years of talking to and learning from teachers, we knew that our first book and a few Thingiverse forums were a good start but would not be enough. The community was calling for high quality lesson plans and tips on best practices from their peers. So we tapped industry- leading educators to help us build Thingiverse Education, the largest collection of classroom-ready 3D printing lesson plans, created by teachers and vetted by MakerBot's education experts.

We were thrilled to share what we had learned in the process and set out to write a new book with the help of the education community. MakerBot combined this new class of educators' unique wisdom with our decade of experience building printers to bring you this; the definitive guide to 3D printing in the classroom.

Introducing, the MakerBot Educators Guidebook.

HOW TO USE THIS BOOK

START WITH THE GUIDEBOOK

Part 1: The basics on how 3D printers work and how to use them, including a crash course on 3D design.

Part 2: Nine teacher-tested 3D printing lesson plans to integrate the technology into the classroom, bringing STEM and project based learning to a variety of subjects and grades.

Part 3: Next steps for building your own lesson plans and going further with prints.

EXPLORE MAKERBOT'S THINGIVERSE EDUCATION™

The nine projects here are just a few of the hundreds available at thingiverse.com/education, where you can filter by subject, grade, or standard to find the perfect project.

Use **Thingiverse Education** and the **MakerBot Educators Guidebook** as inspiration to remix existing lesson plans, create your own, and explore new subjects and applications for 3D printing. Think outside the box!

BECOME A MAKERBOT EDUCATOR™

Are you already using MakerBot 3D printers in your classroom? Join the growing MakerBot Educators community and connect to other teachers, participate in exclusive classroom challenges, and get special gear, discounts, and training.

Apply to become a member at makerbot.com/educators

Have questions or suggestions? Please reach out to the MakerBot Education team: **guidebook@makerbot.com**

PRIVACY AND YOUR STUDENTS

MakerBot takes the privacy of our educator community very seriously. Therefore, we want to be as transparent as possible. By using some of the online resources referenced in this Guidebook, you are consenting to provide MakerBot with a wide range of personal information, as we stipulate in the MakerBot Privacy Policy, which can be found at makerbot.com/legal/privacy

Please make sure that your use of this Guidebook and the online resources referenced inside comply with the privacy standards of your school and school district for students under the age of 13.

For more information, the Federal Trade Commissions (FTC) has set forth very helpful best practices and guidelines for educators with respect to online student privacy; please see ftc.gov/tips-advice/business-center/guidance/complying-coppa-frequently-asked-questions#Sc

TABLE OF CONTENTS

Part 01:
All About 3D Printing

P 10. Chapter One
INTRODUCTION TO 3D PRINTING

P 20. Chapter Two
3D PRINTING IN THE CLASSROOM

P 32. Chapter Three
THINGIVERSE AND 3D MODELING

Part 02:
3D Printing Projects

P 48. Project 01
CLOUD TYPES AND DISPLAY STANDS

P 60. Project 02
3D MUSICAL SHAPES

P 68. Project 03
GO FROM CODE TO CAD

P 80. Project 04
3D PRINTED MINECRAFT CASTLE TEACHES PERIMETER AND AREA

Part 03:

Going Further

P 90. Project 05
THE SPEEDY ARCHITECT

P 106. Project 06
H_2 MAKE IT GO

P 118. Project 07
WEATHER SURVIVIAL CHALLENGE

P 128. Project 08
RUBBER BAND GLIDERS

P 140. Project 09
ARCHIMEDES SCREW BONANZA

P 162. Post-Processing
SANDING 3D PRINTS

P 166. Post-Processing
GLUING 3D PRINTS

P 172. Post-Processing
PAINTING 3D PRINTS

P 178. Post-Processing
SILICONE MOLDING WITH 3D PRINTED MASTERS

P 188. Post-Processing
SILICONE MOLDING WITH 3D PRINTED MOLDS

PART 01:

ALL ABOUT 3D PRINTING

CHAPTER ONE

INTRODUCTION TO 3D PRINTING

MAKERBOT REPLICATOR+ 3D PRINTER

There are several kinds of 3D printing technology. Throughout this book we will focus on the most common type, found in all MakerBot® 3D Printers. Let's dive in and learn the basics.

TERMINOLOGY

FDM®: Fused Deposition Modeling, the 3D printing technology used by MakerBot

Slicing: Turning a 3D model into 2D layers used for 3D printing

Filament: Material used to build 3D printed parts

Extruder: The "hot glue gun" of your 3D printer; it uses filament to draw out the layers of 3D printed parts

Build Plate: Surface on which 3D prints are built

Gantry: Moves the carriage in the x-axis and y-axis

Carriage: Carries the extruder

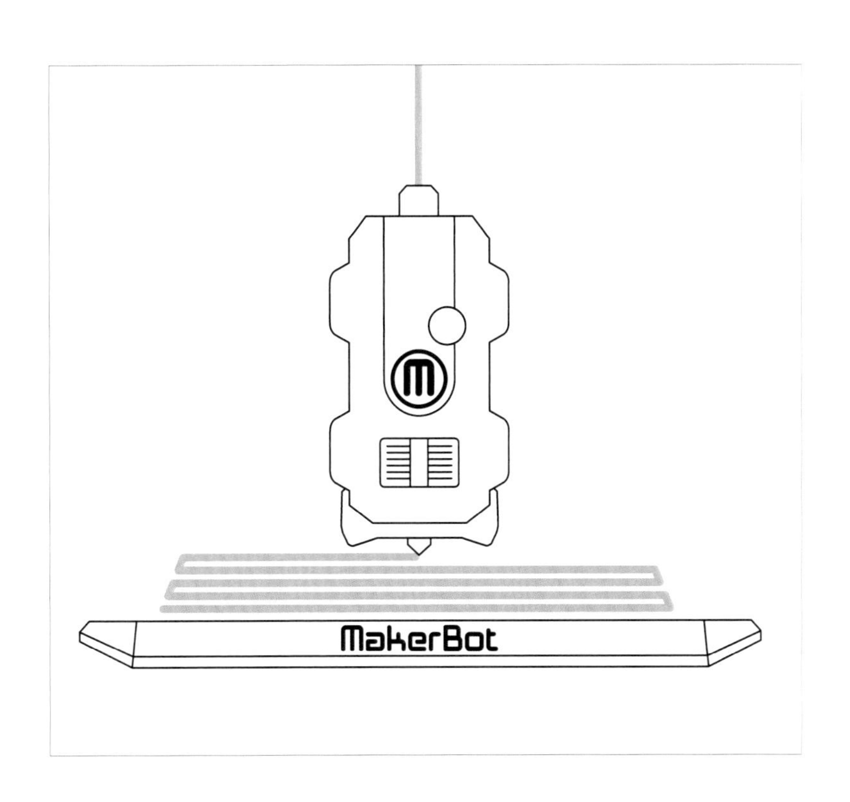

HOW DOES IT WORK?

There are several types of 3D printing technology in use today. The additive manufacturing technology that MakerBot 3D Printers use is called Fused Deposition Modeling, or FDM for short.

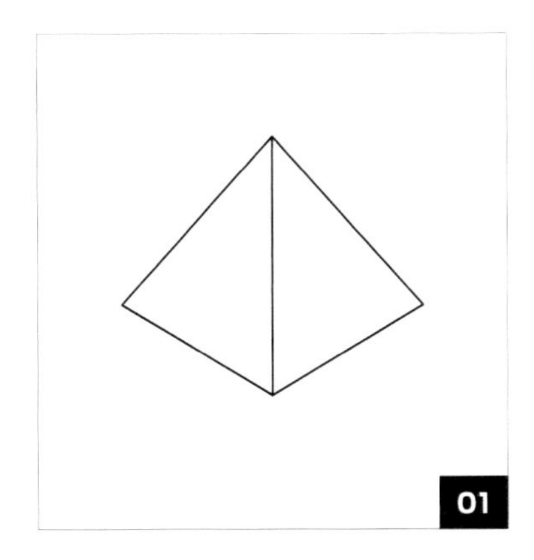

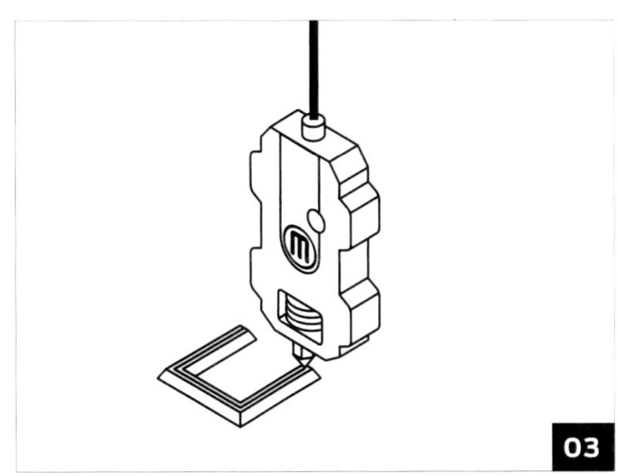

01: FDM printing starts with a digital 3D model, most often generated from a 3D modeling program.

02: The 3D model is sliced into 2D layers using a slicing software and then sent to the printer.

03: On the printer, filament is fed into an extruder that draws out each slice, layer by layer, onto the build plate. Over time, these 2D layers stack on top of each other to build a 3D print.

GET TO KNOW YOUR 3D PRINTER

The following diagrams detail the main components of each of the MakerBot® Replicator® 3D printers. Refer to the diagrams for guidance on major printer components like the build plate, gantry, extruder, and more.

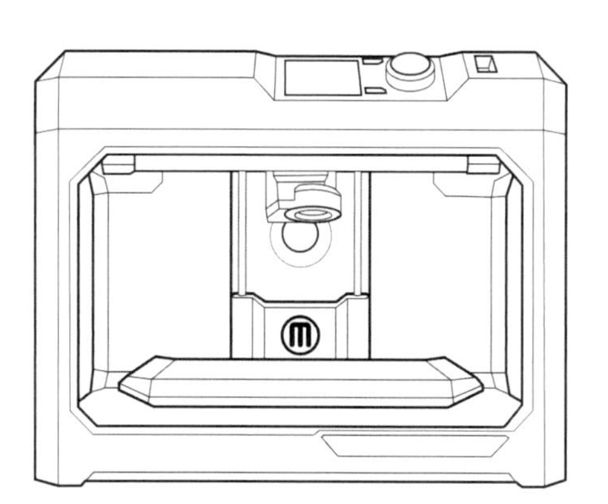

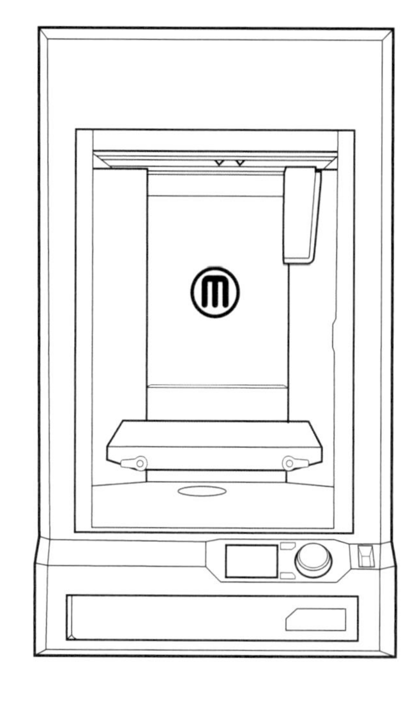

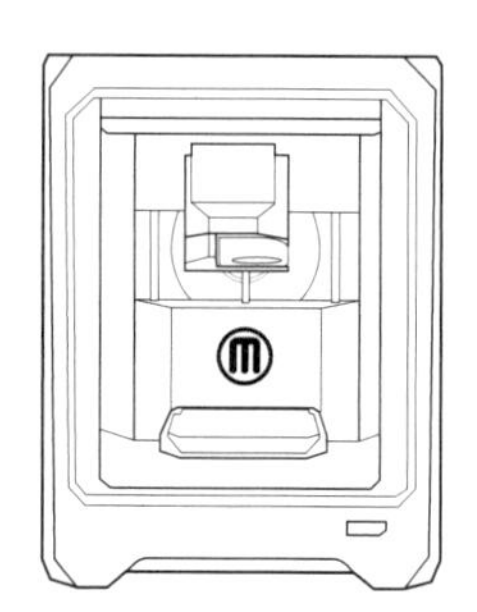

CONTROL PANEL

You can operate the printer directly using the control panel; change filament, calibrate the printer, access printer settings, view wifi status, and more.

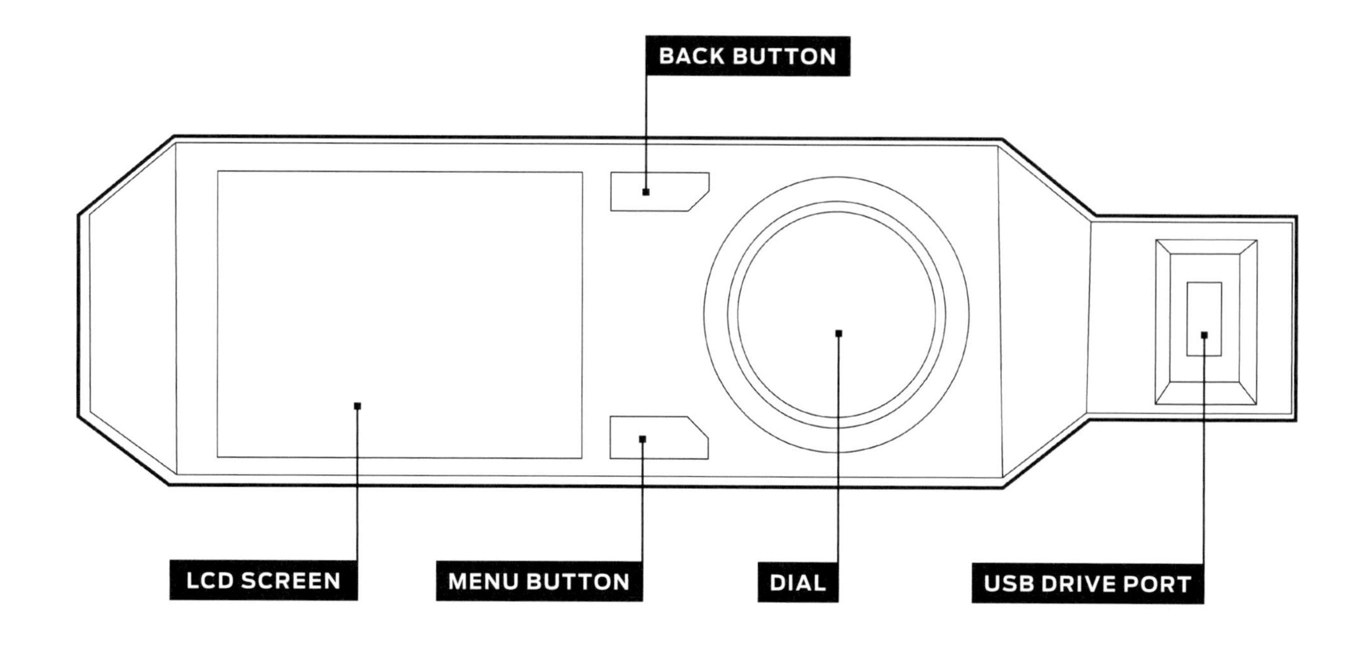

MAKERBOT REPLICATOR+

A favorite amongst educators! The build volume is large enough to print multiple student projects at once and the printer fits nicely on a table or desk.

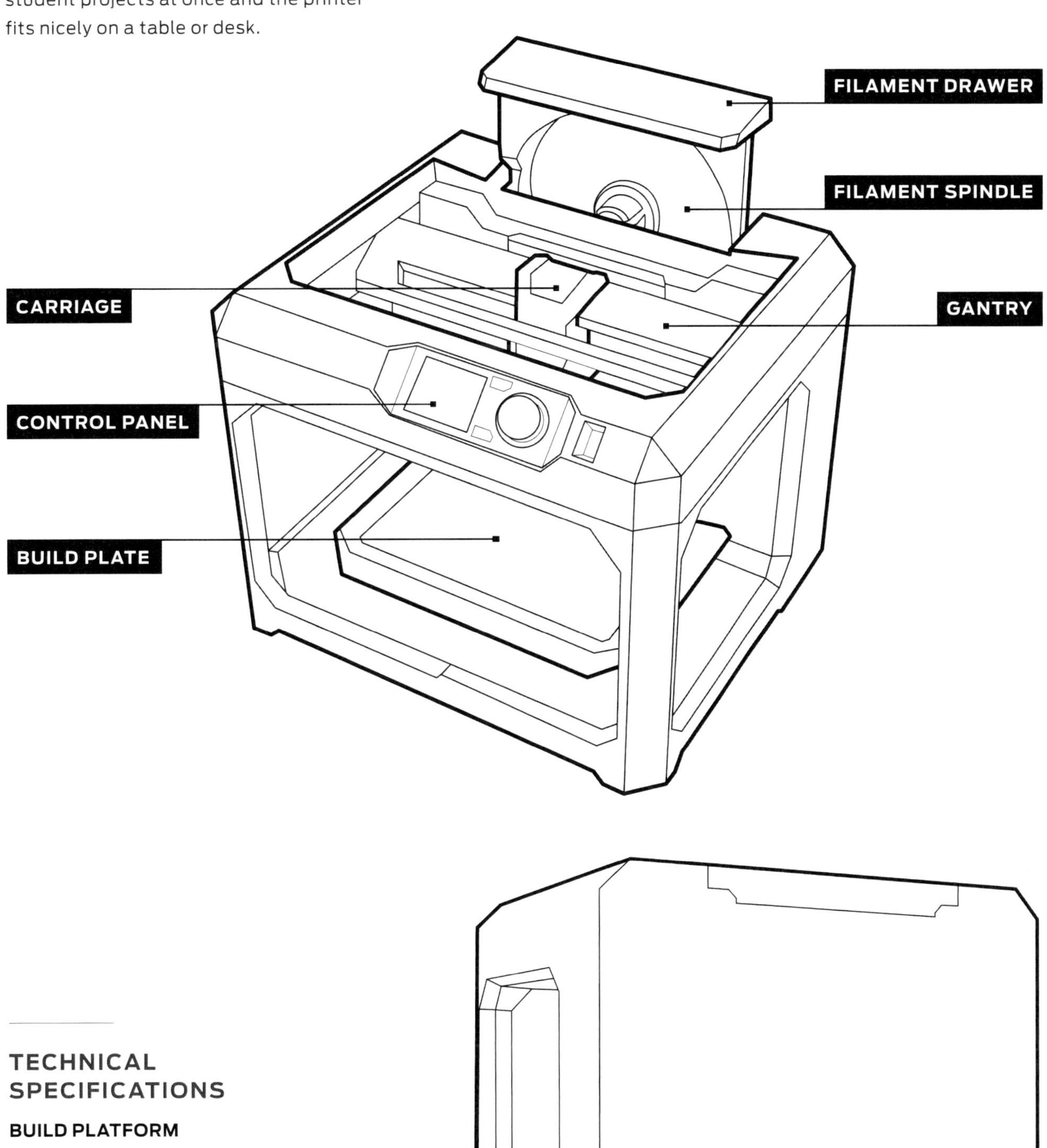

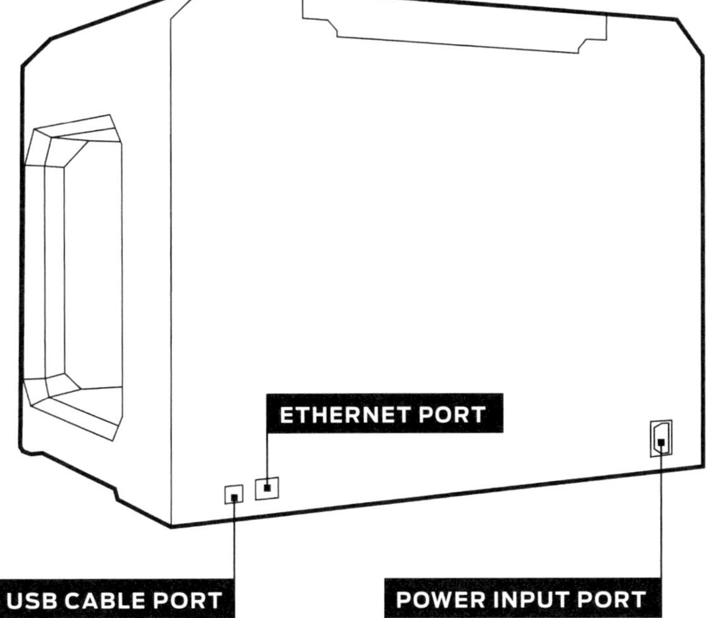

TECHNICAL SPECIFICATIONS

BUILD PLATFORM
Factory-Leveled
Flex Build Plate with
Grip Surface

BUILD VOLUME
29.5 L X 19.5 W X 16.5 H cm
11.6 L x 7.6 W x 6.5 H in

MIN. LAYER HEIGHT
100 microns (0.1 mm)

MAKERBOT REPLICATOR MINI+

Small but mighty! The MakerBot Replicator Mini+ can be easily moved from classroom to classroom.

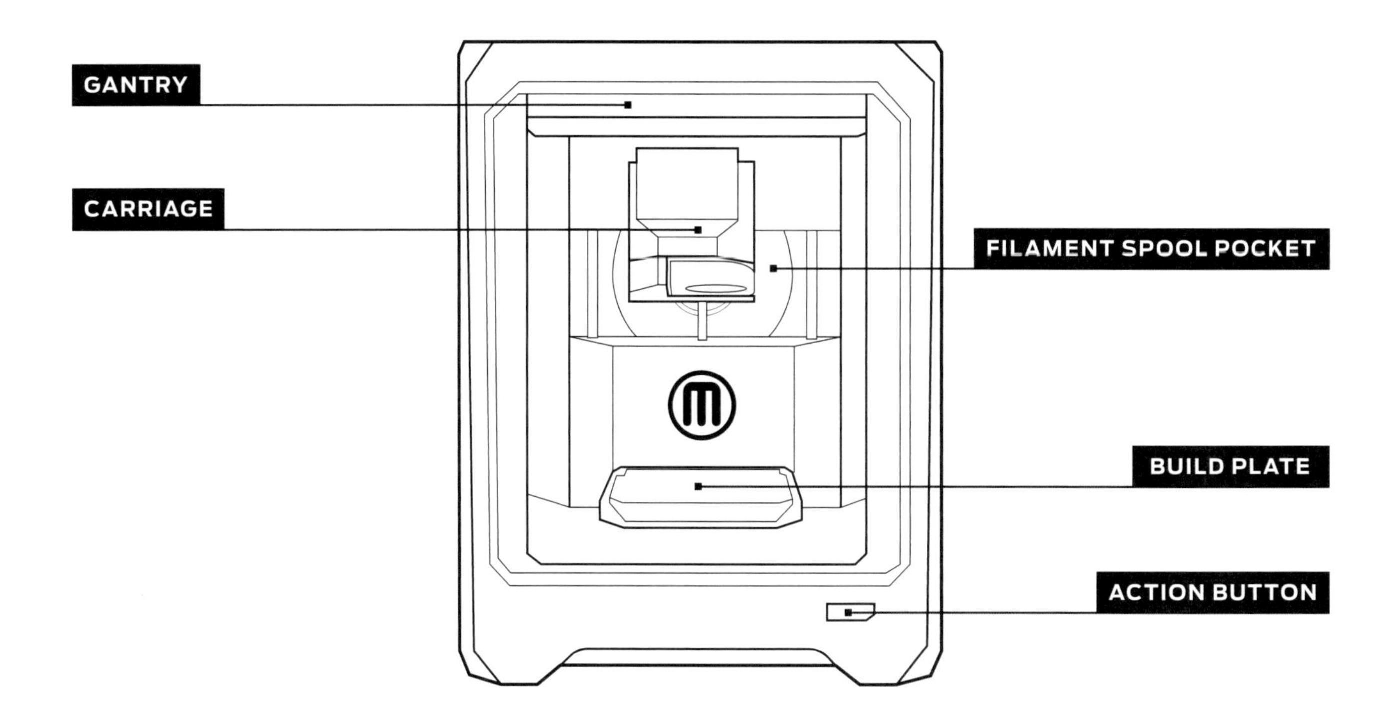

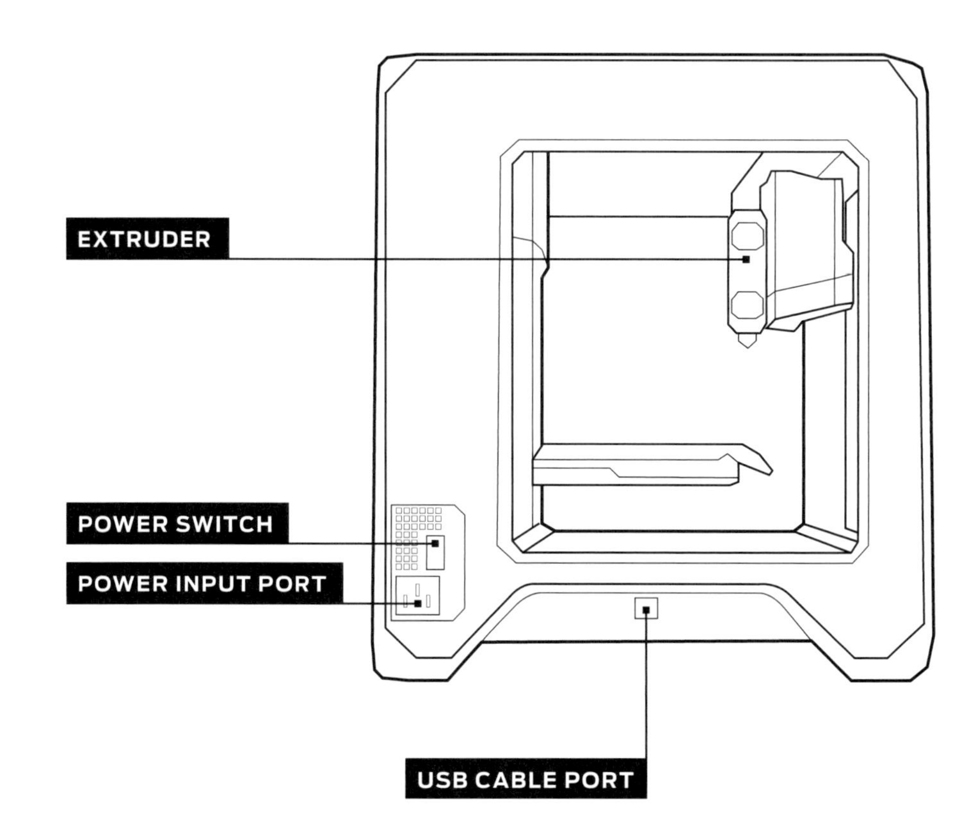

TECHNICAL SPECIFICATIONS

BUILD PLATFORM
Factory-Leveled Build Plate with Grip Surface

BUILD VOLUME
10.1 L X 12.6 W X 12.6 H cm
4 L x 5 W x 5 H in

MIN. LAYER HEIGHT
100 microns (0.1 mm)

MAKERBOT REPLICATOR Z18

The big one! The MakerBot Replicator Z18 is capable of printing parts up to 18" tall, great for large scale projects.

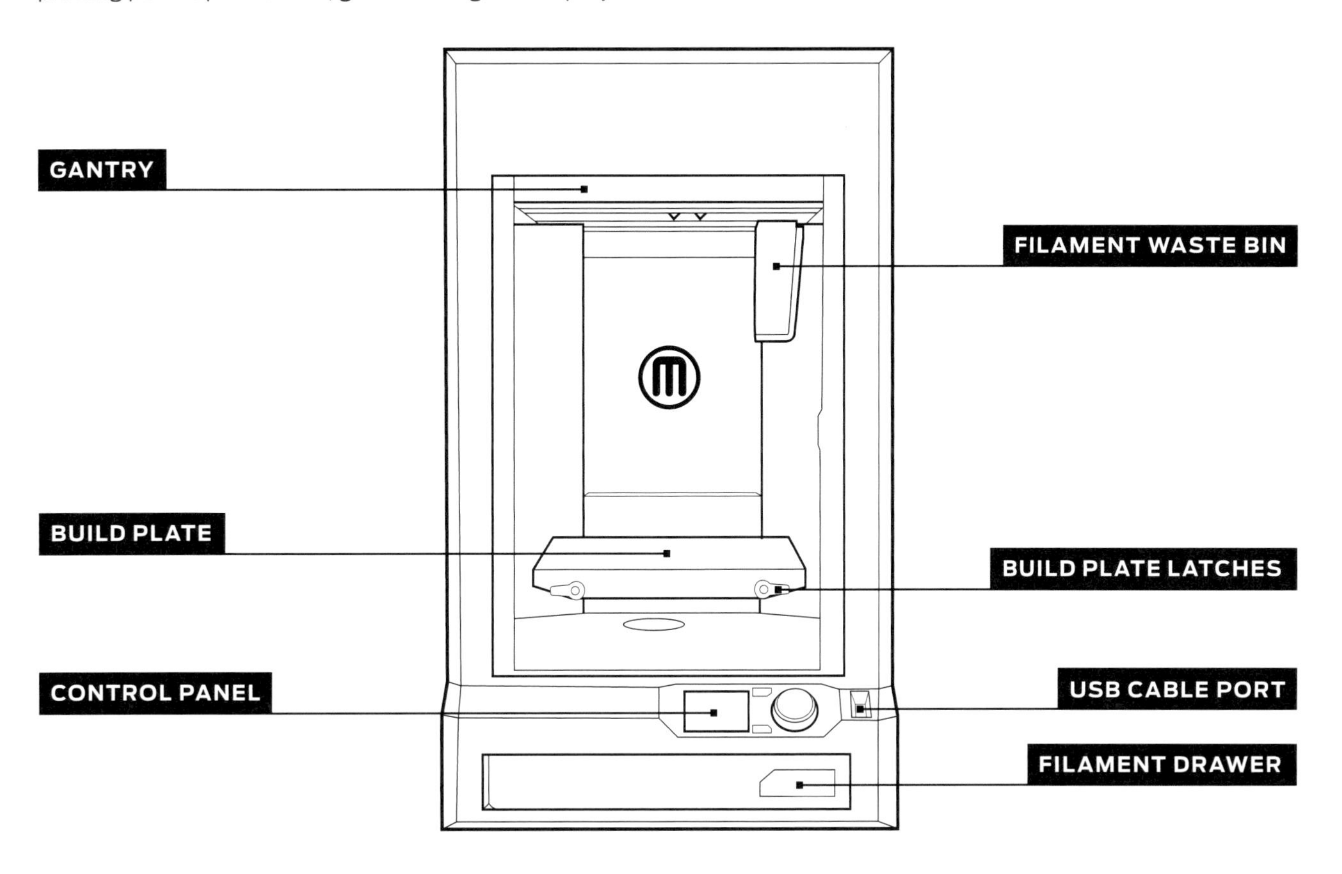

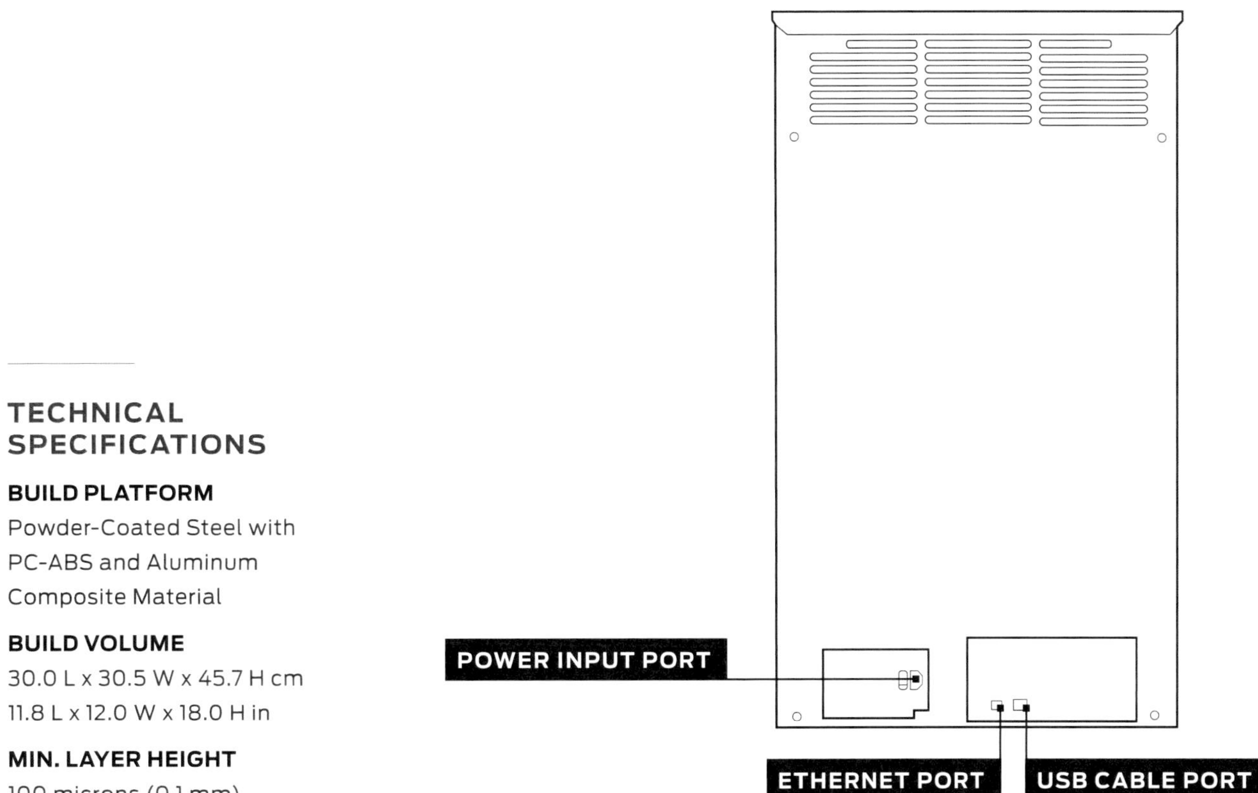

TECHNICAL SPECIFICATIONS

BUILD PLATFORM
Powder-Coated Steel with PC-ABS and Aluminum Composite Material

BUILD VOLUME
30.0 L x 30.5 W x 45.7 H cm
11.8 L x 12.0 W x 18.0 H in

MIN. LAYER HEIGHT
100 microns (0.1 mm)

WHO USES 3D PRINTING?

3D printing is used to solve problems in a variety of industries including engineering, product design, technology, medical, architecture, and even entertainment. Some of our favorite examples include:

Engineering: Engineers at Lockheed Martin use MakerBot 3D printers to fundamentally redesign space telescopes.

Product Design: Canary develops, tests, and refines smart security systems using 3D printing.

Technology: Designers at Peloton use MakerBot 3D printers to prototype their cutting edge indoor bikes.

Medical: Open source 3D printable prosthetics allow people all over the world to print low-cost assistive devices.

Architecture: Perkins + Will, an architecture firm in New York City, uses 3D printing to test new building ideas.

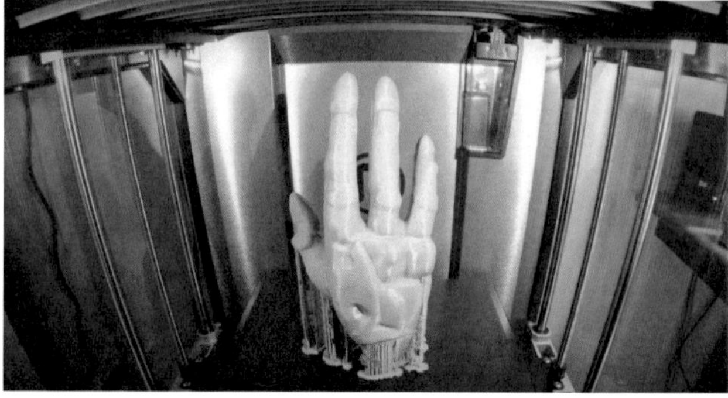

Entertainment: The Legacy Effects team created a full-scale, detailed alien suit using 3D printing.

HOW TO USE A 3D PRINTER

These are the 3 basic steps to print on a MakerBot 3D printer.

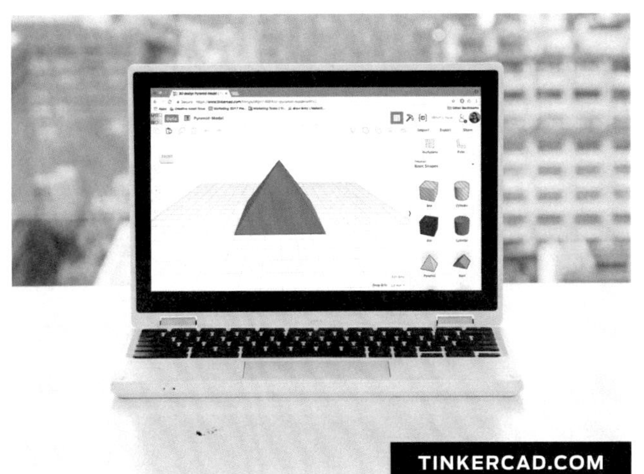

1: Design. In order to 3D print, you must start with a 3D file. Here are a few ways to get one:

› **Design** a model to print in a 3D design software or computer-aided design (CAD) program.

› **Scan** an existing physical object with a 3D scanner.

› **Find** a model online from websites like thingiverse.com or grabcad.com

2: Slice. Before printing a model, prepare the file in MakerBot® Print™. Follow these steps:

› **Edit** the print settings.

› **Decide** if you want to print more than one part.

› **Slice** the model to prepare for printing, which translates the model(s) into a language the 3D printer can understand.

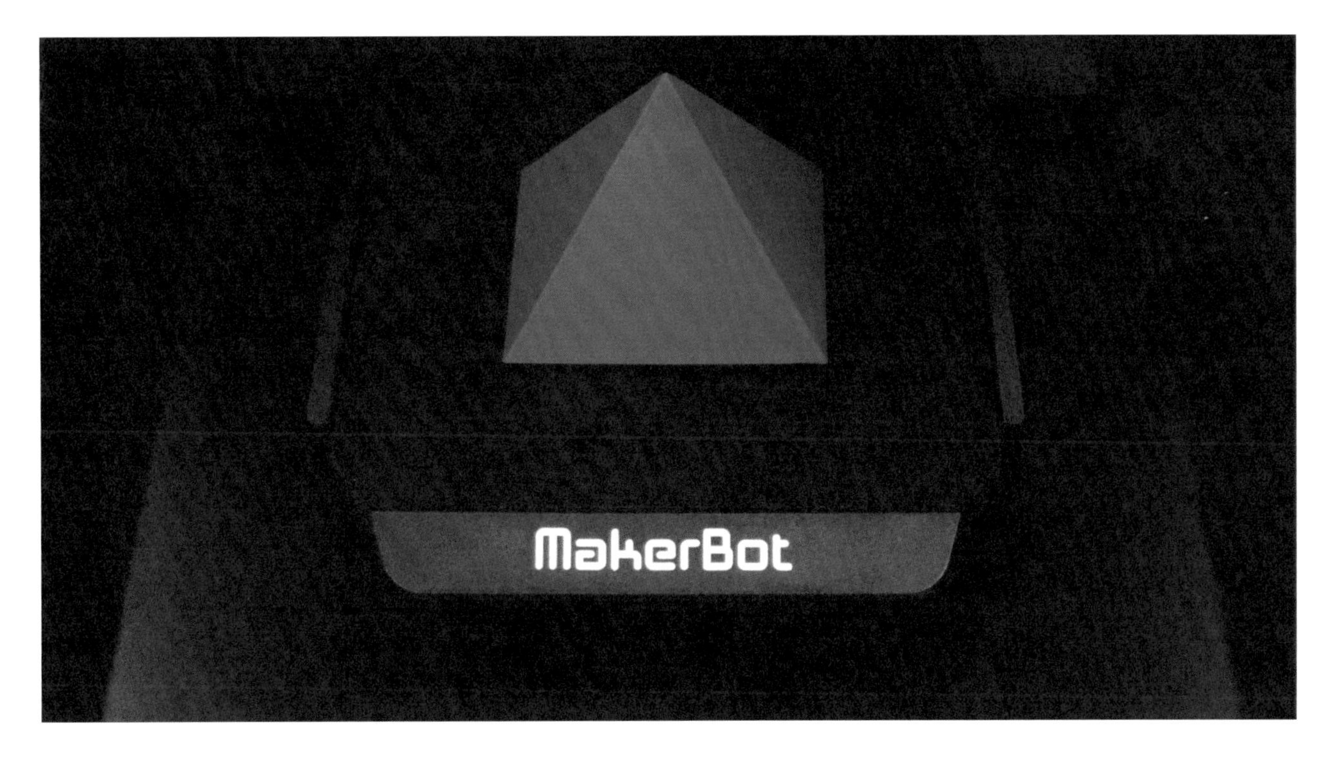

3: Print. Send your sliced file to your MakerBot 3D printer for printing.
The print time will depend on many factors, including:

› Your print settings

› The size and complexity of the model

To learn what 3D printer is right for you or to
speak directly to an implementation expert, visit:

MakerBot.com/learnmore

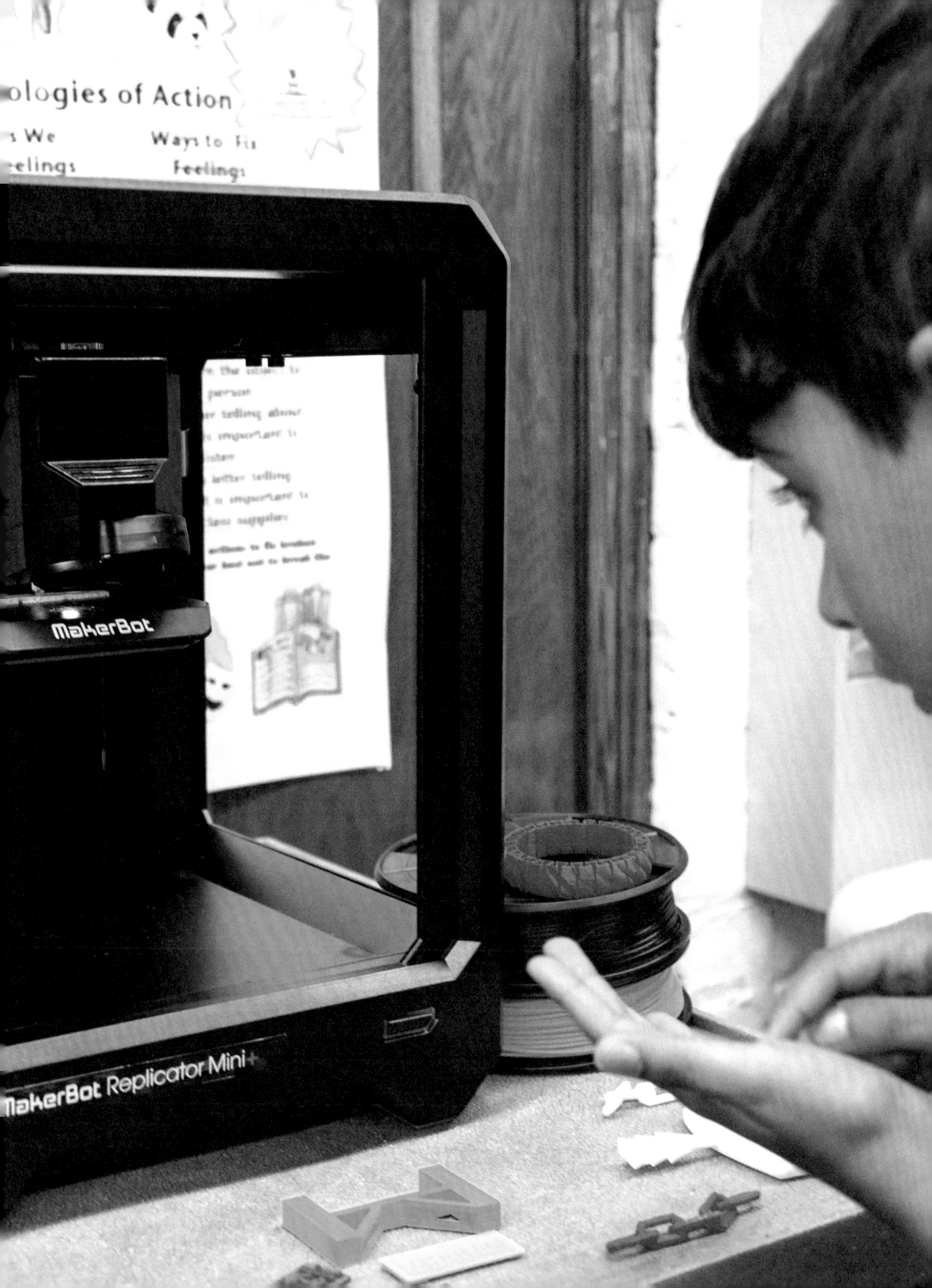
ologies of Action
Feelings
MakerBot
MakerBot Replicator Mini+

CHAPTER TWO

3D PRINTING IN THE CLASSROOM

CYPRESS HILLS COMMUNITY SCHOOL *Brooklyn, New York*

Now that we've explored how others are using their 3D printers, it's time for you to get started. This section will provide advice for setting up 3D printers in your school, classroom, library, or makerspace.

TERMINOLOGY

Smart Extruder+: Extrudes filament to draw out the layers of your 3D prints. The "hot glue gun" of your 3D printer

Purge Line: Straight line drawn across the front of the build plate at the start of every print

MakerBot Print™: MakerBot's free software for managing and preparing your 3D print files

Print Preview: Gives important time and material estimates

Raft: Flat surface that provides a large foundation for print adhesion

Support Material: Removable scaffolding structure that is built underneath overhangs (unsupported sections) of your printed parts

Infill: Support structure built inside of the printed parts, measured in density

Shells: Outside walls that make up the perimeter of printed parts

Layer Height: The height of each layer of a printed part

HOW OFTEN DO YOU USE YOUR PRINTER?

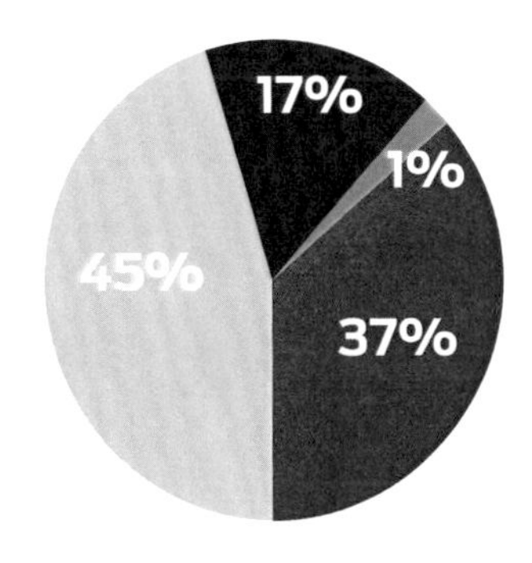

Weekly
Daily
Monthly
Never
Less than once per month

SCHOOLS WITH MAKERBOT

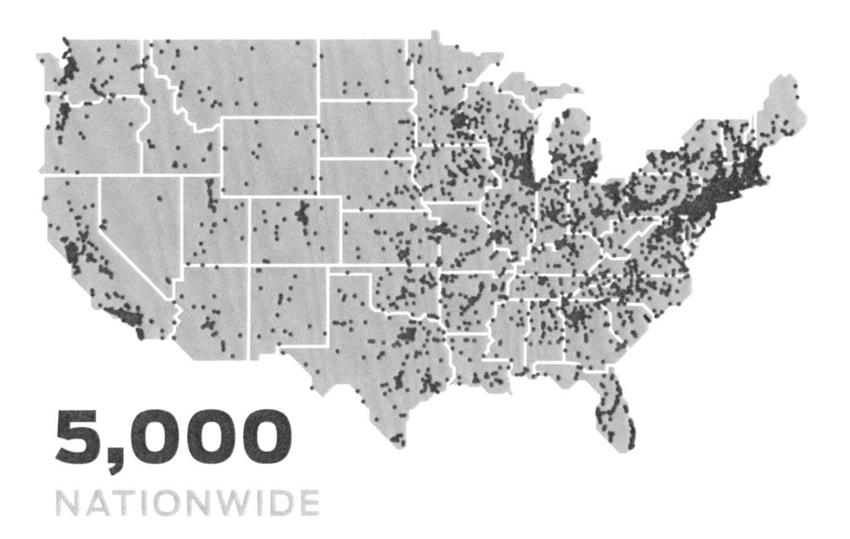

5,000
NATIONWIDE

MakerBot printers are reliable, safe and easy to use within the classroom. They are the #1 choice amongst educators in over 5,000 schools nationwide. MakerBot products helps to prepare students for meaningful careers by opening up a world of design thinking and student innovation.

WHAT SUBJECTS DO YOU USE YOUR 3D PRINTERS WITH?

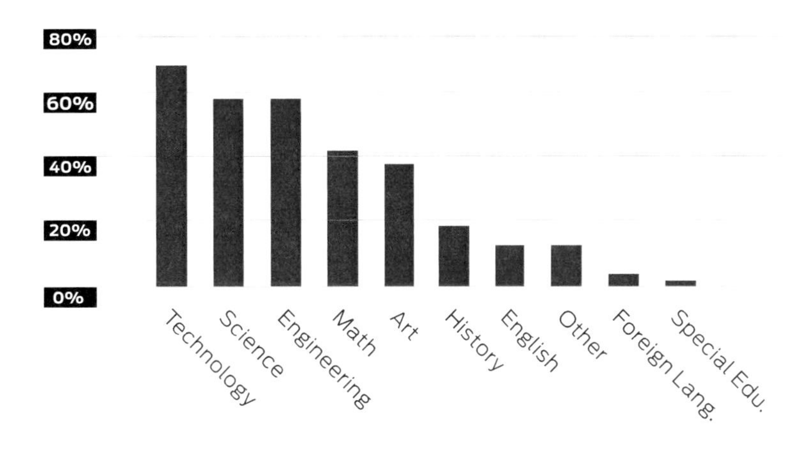

MakerBot Thingiverse Education provides teachers with easy-to-implement lesson plans written by our growing community of MakerBot Educators. These lessons contain detailed STEM projects aligned to Next Generation Science Standards (NGSS) and Common Core State Standards (CCSS).

PREPARE YOUR CLASSROOM

Your MakerBot Replicator+ is versatile and can be used almost anywhere in the classroom. With that in mind, the placement of your printer(s) can have an impact on its use as well as its performance. 3D printers are best used in a relatively temperature stable environment, where there are no frequent gusts of wind, dust, water, or temperature changes.

MakerSpace Setup: Students work side by side with their printers and peers for enhanced collaborative learning.

General Printing Area: Offers students a dedicated space to house communal printers.

Instructor's Printer: Allow students to follow detailed instructions as they prepare for their printing experience.

SUGGESTED 3D PRINTING TOOLS AND SUPPLIES

USB STICK
stores and organizes print files and documents

FLUSH CUTTERS
removes support material

NEEDLE NOSE PLIERS
removes support material

CRAFT SPATULA
removes prints from the build plate

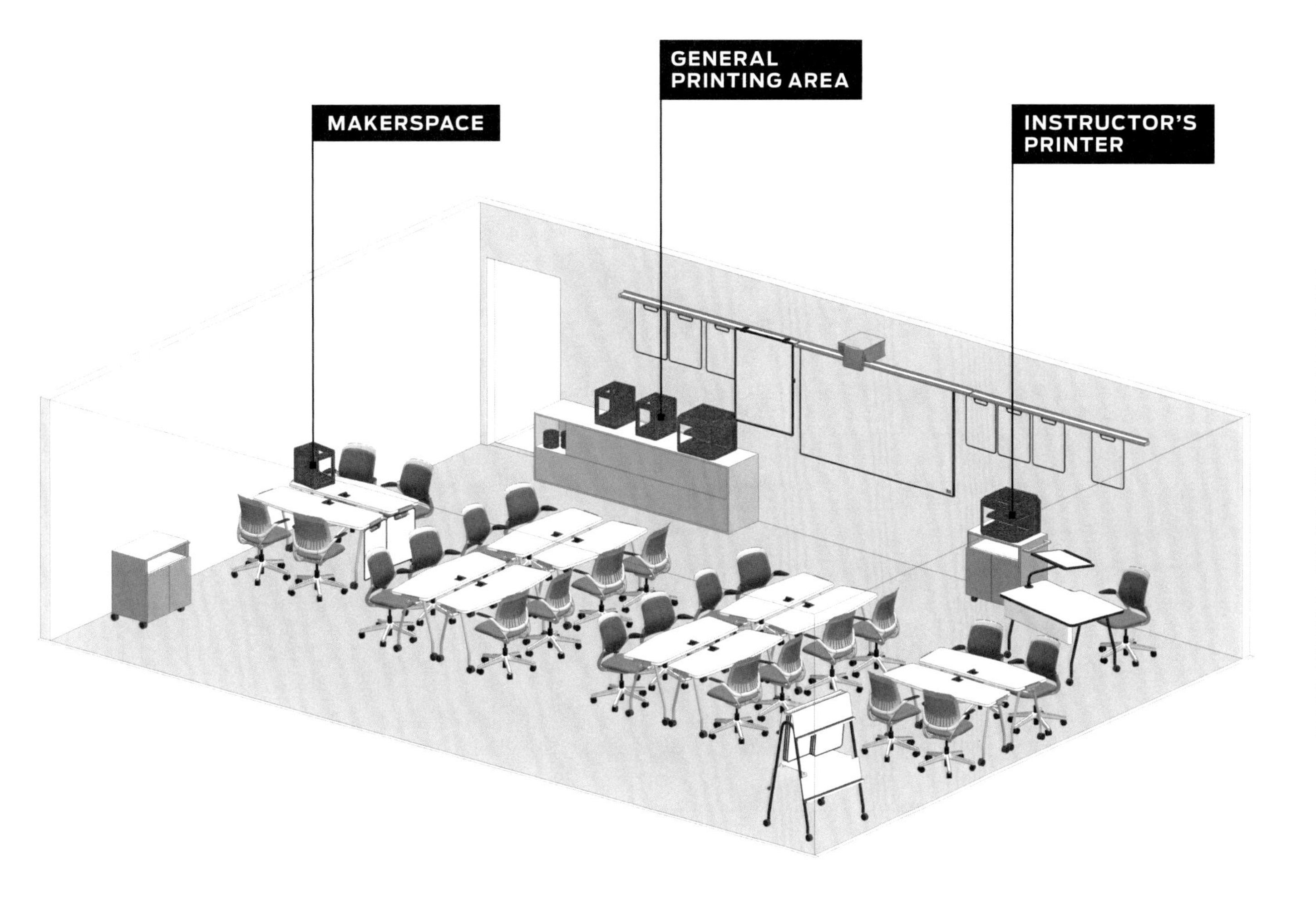

PROS AND CONS OF PRINTER PLACEMENT IN YOUR SCHOOL

IN THE CLASSROOM:

Pros: Allows for lots of student interaction, hands-on time, and printing in class

Cons: Lessens access to the rest of the school

IN A COMMUNAL SPACE:

Pros: Ensures visibility, encourages more student, faculty, and staff use

Cons: Makes it more difficult to print during class

DO'S

› Place your 3D printer on a stable surface

› Have a small workspace near your printer

› Keep print tools handy

› Store filament in a cool, dry location, ideally in its original packaging

DON'TS

› Place your 3D printer near air conditioning

› Store filament and extra Smart Extruder+'s in unsecure areas

PRINTER UNBOXING AND SETUP

When your printer arrives, the first thing you'll need to do is unbox it and begin the setup. You'll be guided to connect to WiFi, attach your Smart Extruder+, calibrate, load filament, and begin a test print. The setup process should take 15-20 minutes and then you're ready to set off your first 3D Print!

For a comprehensive guide on how to unbox your MakerBot 3D printer, visit *makerbot.com/getstarted*

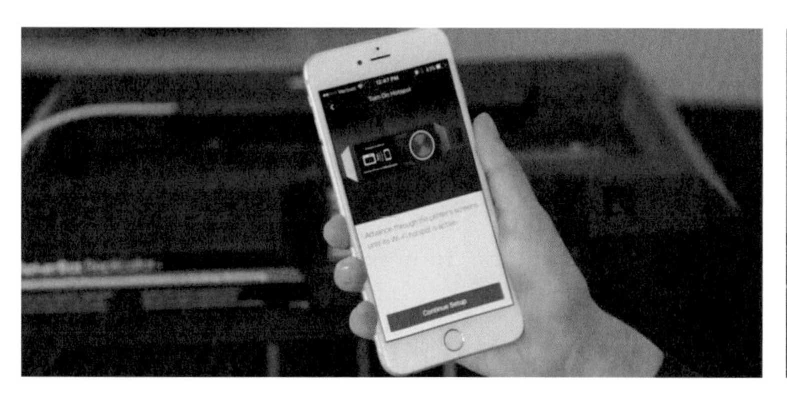

WiFi and Remote Monitoring

During the setup you'll be asked to connect your printer to a WiFi network. Connecting to WiFi allows you to start and monitor prints remotely from your printer's on-board camera.

If your school has restricted or inconsistent WiFi access, you can skip this step. You'll still be able to connect to the printer via USB cable, USB stick, or Ethernet cable.

TIP: Save all your packing materials. You might need to move or ship your printer at a later date.

RUNNING A TEST PRINT

The last step in the guided setup process is to select a test print from the printer's internal storage. Here are some things to look for when printing:

Purge line: Your MakerBot Replicator+ draws a straight line across the front of the build plate at the start of every print. If you have a MakerBot Replicator Mini+, this will be on the side.

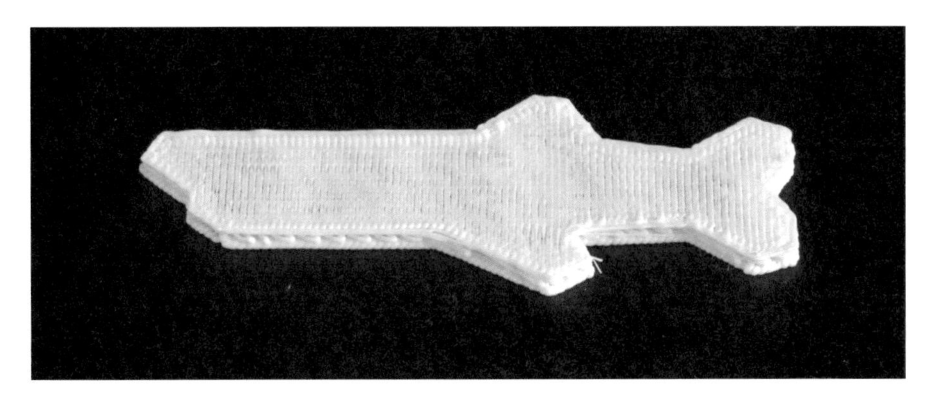

First layer: The first layer of your print is the most important. If you notice that parts of the first layer are not adhering to the build plate, cancel the print and consider leveling your build plate.

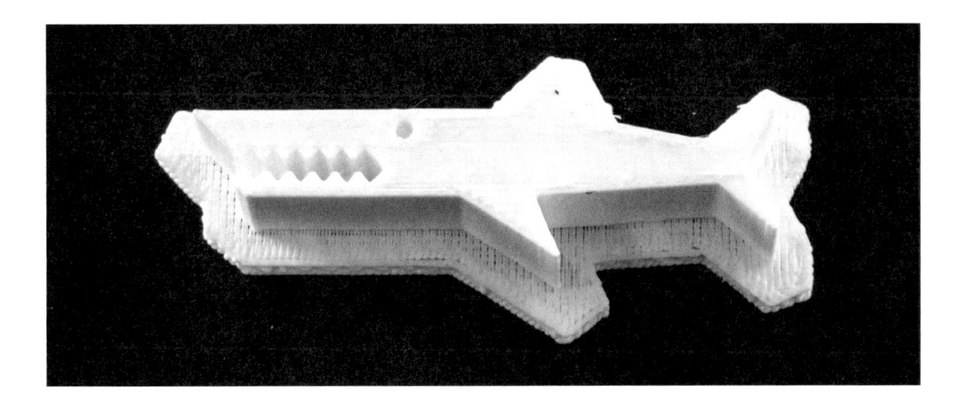

Test Print: Each of the test prints on your printer take between 20-45 minutes to print. Once the test is complete, you're ready to move on and begin printing!

Mr. Jaws *@Mahoney*
Thing: #14702

PREPARE FILES FOR PRINTING WITH MAKERBOT PRINT

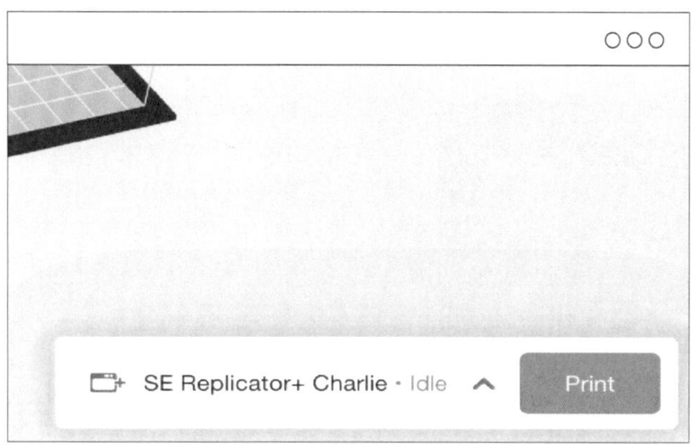

Select Printer: Click on the **printer menu** to view your active printers. Select **add a printer** to add a new printer to your list.

Add a network printer to browse from printers already on your network via WiFi or Ethernet.

Connect via IP address to add a printer using its IP, this can be found on the MakerBot Replicator+ onscreen menu. **Add an unconnected printer** if you plan on transferring files to your printer via USB stick.

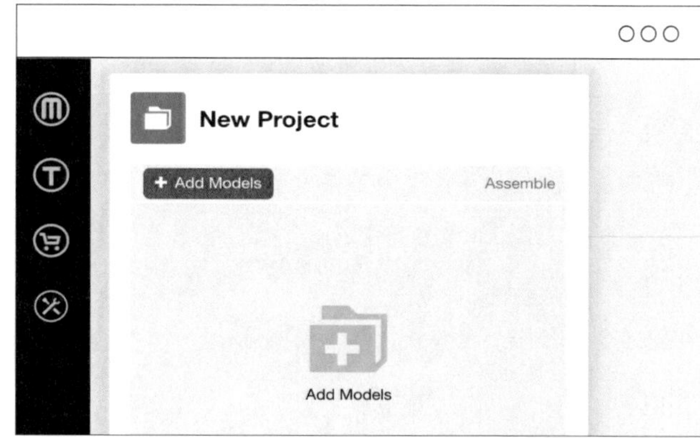

Insert Files: Open the project panel and select **add models**. Alternatively, you can drag and drop files directly into MakerBot Print's main window from your computer.

File Types: Both Mac and PC users can import .STL files. PC users can import native CAD files from programs like SolidWorks® and Autodesk Inventor®.

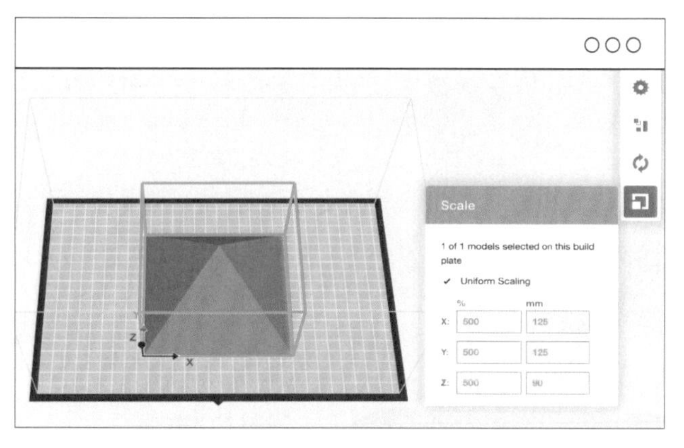

Layout: Arrange models on your build plate(s) using the **arrange**, **orient**, and **scale** menus. Rotate your models so that the largest flat surface is touching the build plate. Try using **place face on build plate** in the **orient** menu to help.

TIP: Place models as close to the center of the build plate as possible. Group models as close together as possible without overlapping, **arrange build plate** is useful for arranging lots of models.

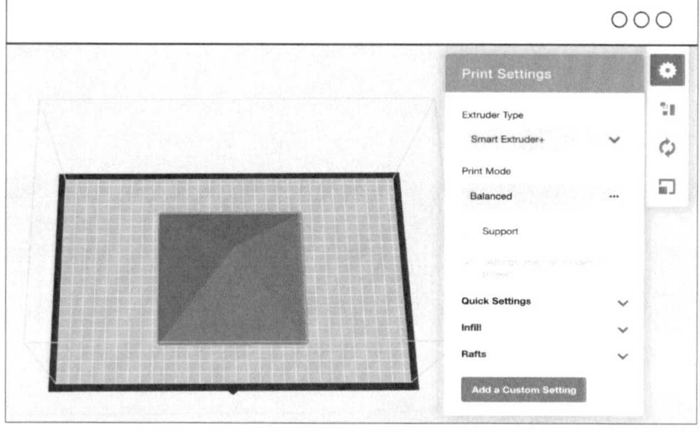

Settings: Select your print mode, extruder type, and toggle support material on/off. Print settings will affect the strength, surface quality, weight, print time, and other properties of your printed parts. We recommend always printing with rafts on.

TIP: Print modes are a set of recommended print settings. For more advanced control, select the **add a custom setting** button to create your own custom print modes.

Once you have your printer set up, it's time to learn how to prepare files for printing. MakerBot Print is the software you'll use to take 3D designs and convert them to 3D printable files. Download it for free at makerbot.com/print

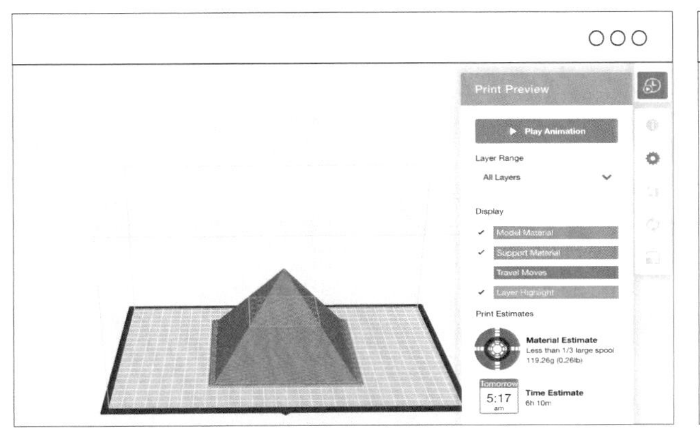

Print Preview: Once you have your models set and have selected your print settings, click the **estimates and print preview** button to see how your models will be printed and to estimate print time and material usage.

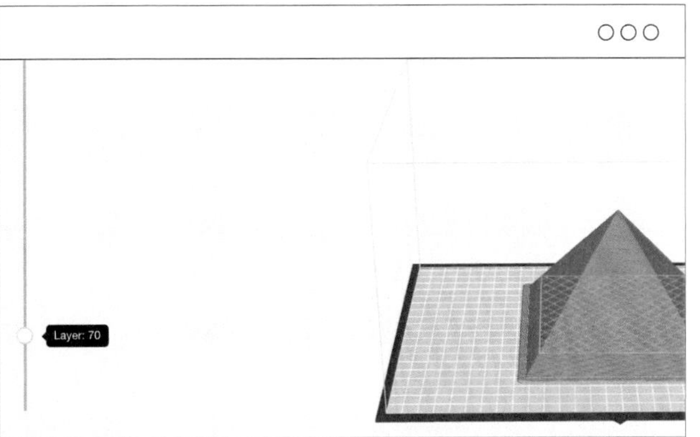

TIPS: Use the scroll bar on the left to show students how parts will be printed.

Some models with large overhangs and bridges will need supports. Toggle supports on and off then **preview** to learn where supports are generated.

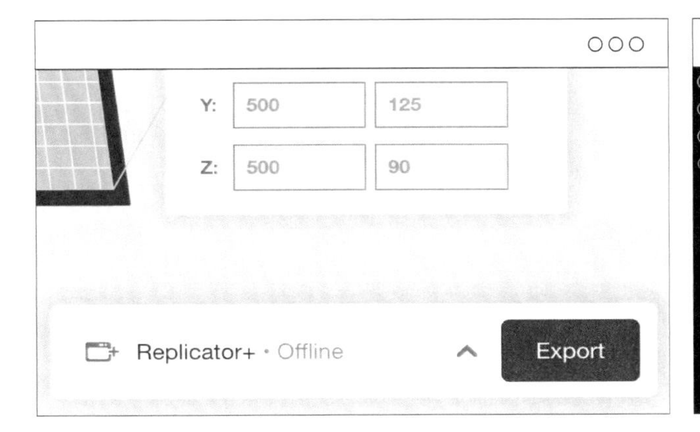

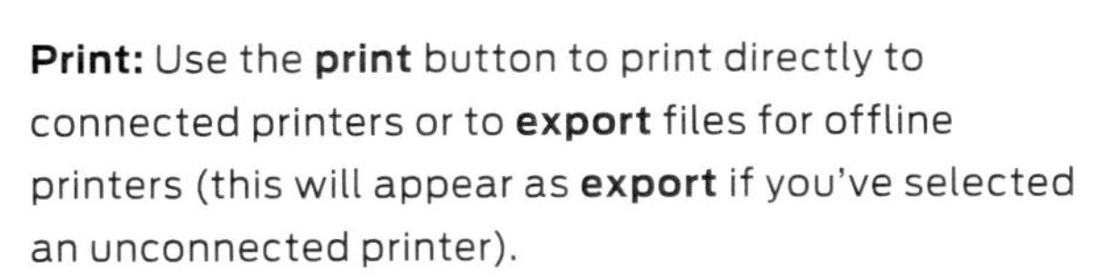

Print: Use the **print** button to print directly to connected printers or to **export** files for offline printers (this will appear as **export** if you've selected an unconnected printer).

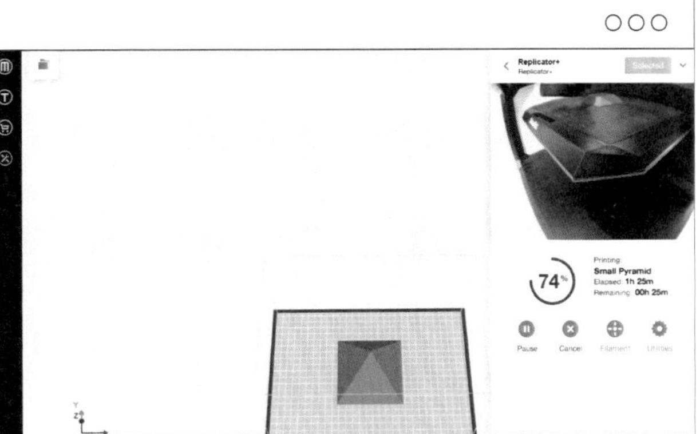

Monitor your print progress by navigating to the **printer menu** and watching the live camera feed. You can also monitor via the MakerBot® Mobile™ app.

SHELLS AND INFILL

Print settings can dramatically change the strength, appearance, print time, and other properties of your printed parts.

Shells are the perimeter on each layer; they make up the walls of your part. Infill is the internal structure of your part. You can set the infill of your part to be anywhere from 0% (hollow) to 100% (solid). Increasing the infill and number of shells will make your parts stronger, but will increase print time and filament use.

PRINTED PYRAMID
10% Infill / 02 Shells
without supports

PRINTED PYRAMID
0% Infill
08 Shells

PRINTED PYRAMID
25% Infill
02 Shells

PRINTED PYRAMID
02% Infill
02 Shells

PRINTED PYRAMID
50% Infill
02 Shells

SUPPORTS AND RAFTS

Supports are printed scaffolding for overhangs. If your model has overhangs greater than 68 degrees (measured from the vertical axis) then you will need to print with supports. A raft helps the part adhere to the build plate by laying down an even, flat foundation to print on.

3D Model: The T model has overhangs greater than 68 degrees and needs support material. The Y model does not need support material.

Supports: After printing, the T will needs support material removed. Both printed with rafts.

Final Print: Final parts after removing supports and rafts.

PRE-PRINT CHECKLIST

SOFTWARE

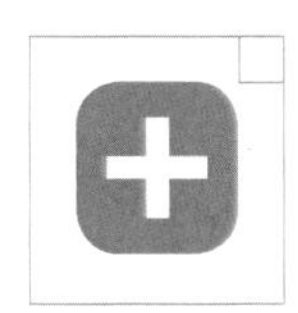

01. Add Files: Click **File** > **Insert File** or drag and drop right onto the build plate

02. Arrange: Organize objects by dragging or using **Arrange Build Plate**

03. Print Settings: Adjust your print settings to change print speed and quality

04. Print Estimates and Preview: Double check print time and material usage by checking **Print Preview**.

HARDWARE

05. Install Build Plate: Load your build plate onto the Z-stage and confirm it's snug

06. Attach MakerBot Smart Extruder+: Confirm that the Smart Extruder+ is attached properly

07. Calibrate MakerBot Smart Extruder+: If you've just attached a Smart Extruder+, run a Z Calibration. On the printer, select **Settings** > **Calibration** > **Calibrate Z Offset**

08. Load Filament: Select **Filament** > **Load Filament**. Then, ensure that the filament is seated in the drawer, fed through the guide tube, and securely inserted into the Smart Extruder+ when prompted.

POST-PRINT CHECKLIST

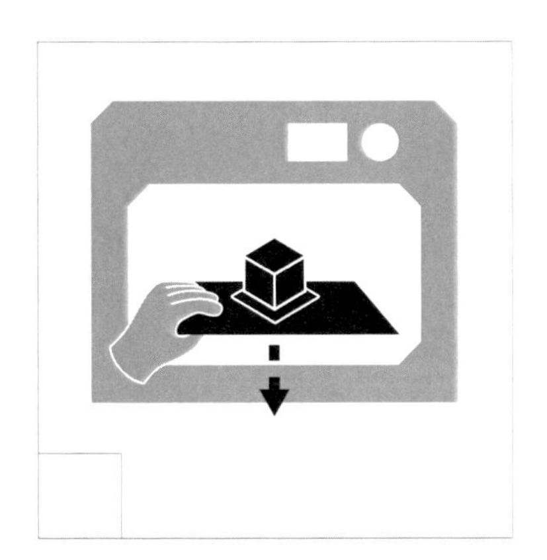

09. Remove Build Plate: Slide the build plate toward you to remove it from the Z-stage

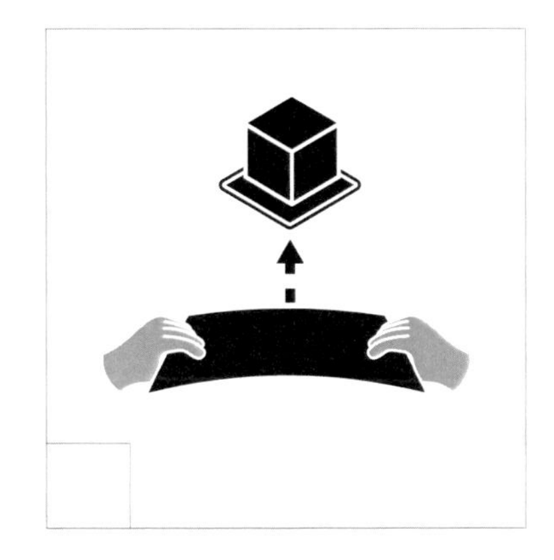

10. Remove Parts from Build Plate: Gently flex the build plate or use a thin craft spatula to remove parts

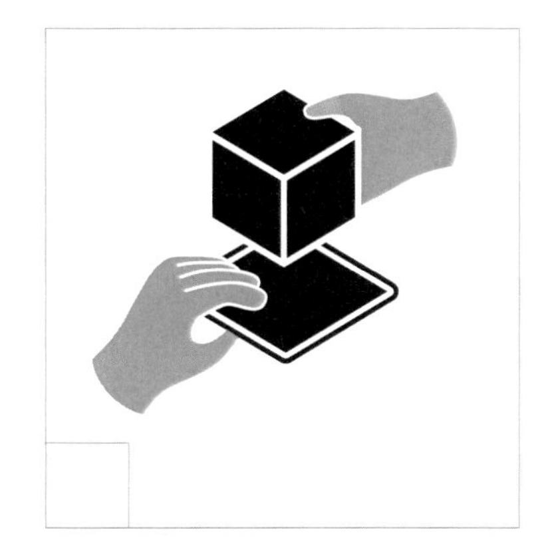

11. Discard Rafts and Support: Use your hands or tools to gently remove the raft and support materials from parts

ADVICE FROM MAKERBOT EDUCATORS

WHAT TO KNOW BEFORE YOU START

We asked educators to tell us one thing they wish they knew before they started using a 3D printer in the classroom:

Combine student projects: "I wish I knew that you could combine multiple student projects into one large print file. This has saved me loads of time printing."

Understand your printer: "Using a printer is very straightforward, but you should allot sufficient time to print some models yourself to really test the abilities and limitations of 3D printing. That way, you'll be better able to look at models and determine if they're printable and whether or not they'll need supports."

Learn how to 3D design: "At first I relied heavily on Thingiverse, which was fun and informative, but it wasn't until I felt comfortable designing alongside my students that I felt like I truly unlocked the teaching potential of my MakerBot 3D printer."

Be open to ideas from students: "Letting them have the 'ah-ha' moment is meaningful."

Mind your filament: "Make sure you have enough filament before starting a project. When storing filament, keep the end tucked into the spool to prevent tangling, and keep the spools in a cool, dry place."

HOW DO YOU ANALYZE AND GRADE STUDENT PROJECTS?

Our educators shared how they assess student success:

Clear, fair requirements: "Your students will all be at different 3D design skill levels, so set requirements that focus on problem solving, collaboration, creative thinking, etc. that are mutually attainable for all students."

Look for the basics: "Grade on basic 3D printing best practices, like ensuring that models are flat against the build plate and that parts have no obvious weak points."

Allow for mistakes: "Create a rubric that allows for mistakes. Overcoming flaws in design is when the real learning occurs; even if their designs fail, if a student can accurately explain why, they've done well."

HOW DO YOU MANAGE STUDENT FILES?

Managing multiple student files can be tricky. Here are some tips from teachers for dealing with student files in your classroom:

Establish a naming convention: "Have students name their files with their Project Name, Student Name, Date, and Version (if applicable). i.e: *Clouds_StudentName_052017_V1.stl*"

Set up student accounts ahead of time: "Create MakerBot and 3D software accounts on behalf of students prior to beginning a project and log in on each computer in advance."

Create communal storage: "There are a variety of options available to manage student files online, including Google Classroom™, Google Drive™, Dropbox®, and more. When WiFi isn't available, use a shared USB stick to keep all student files in one location."

If you have one printer, print in bulk: "Add multiple student models onto one build plate in MakerBot Print and print during after-school hours and weekends. (Just make sure to watch the first layers before leaving.)"

Review student files before printing: "Schedule time to review students' work in the design program, and help them export their models as .STL files. Then, help them import into MakerBot Print, check their settings, and begin printing. Students can also upload their files to Thingiverse."

WHERE THE MAGIC HAPPENS

Grow your own 3D printing skills and inspire your students:

Don't be afraid to try: "I taught myself how to use our 3D printer during class. The kids watched me make mistakes and I think it really taught them a lot about the learning process."

Be a model: "Becoming a 3D designer and printer yourself is the best way to inspire students to do the same. Don't be afraid to play and make mistakes while learning how to design and print. We expect our students to be brave, lifelong learners, and we must reach for that same goal!"

CHAPTER THREE

THINGIVERSE AND 3D MODELING

When exploring how to best use your 3D printer, you don't have to go it alone. There's a massive online community of millions of 3D printing educators and makers that actively offer advice, answer questions, and contribute their designs for others to use freely. This community is Thingiverse.

Thingiverse is built on the principles of sharing, learning, and making. It's a community where users from all over the world can download free 3D models and 3D printing lesson plans for any age group or subject. MakerBot founded Thingiverse in 2008 while developing the first desktop 3D printers, and several years later it's grown to an enormous size and stands in support of the entire 3D printing industry.

TERMINOLOGY

Thingiverse®: The largest online 3D printing community and library of printable files, accessible at thingiverse.com

Thing: A 3D model uploaded to Thingiverse, this can be a single object or several, and typically comes with pictures and printing instructions

Thingiverse Education™: 3D printing projects and lesson plans for different grades and subjects

Remix: A cornerstone of Thingiverse. This is when a user edits or builds upon a Thing to customize it in some way while providing attribution to the original designer

Licensing: Describes how a designer permits the community to use his/her creation (i.e. whether it can be remixed or sold commercially)

Makes: When a user downloads and prints another designer's model, then uploads a picture of it to demonstrate its appearance and printability

Collection: A curated folder of Things by users, and a great way to organize Things for your class

THINGIVERSE, A UNIVERSE OF THINGS

Thingiverse is a great place to go for inspiration when creating your own 3D models. Instead of designing a hinge or spring from scratch, there are plenty of Things you can look to for guidance, or download and modify to suit your own project.

Several designers use Thingiverse the way photographers use Instagram; designers upload 3D models to share with others and get feedback and recognition for their work. Take some time to explore the more active users' portfolios – you're sure to find some amazing and inspirational Things!

Be mindful of the licensing features that the original designer has chosen when downloading and printing 3D models. In some cases, you are free to use the model however you'd like. However in others, you may not be allowed to remix or use the model commercially, such as selling prints of the model. In all cases, you must provide attribution to the designer, especially when you remix another designer's model.

THING BADGES

When searching Thingiverse, keep an eye out for these icons to help you identify standout Things:

Verified Thing: 3D model has been print tested and verified to print successfully.

Featured Thing: This Thing has been featured on the Thingiverse homepage.

Challenge Winner: This Thing was the winning entry in a Thingiverse Challenge.

Thingiverse Education: This Thing is an educational project.

EXAMPLE THINGS

Subaru WRX EJ20 Boxer Engine Model
Thing: #1643878
@EricThePoolBoy

Vases
Thing: #284006
@LeFabShop

Mars Rover
Thing: #404824
@MakerBot

WALLY - Wall Plate
Thing: #47956
@TheNewHobbyist

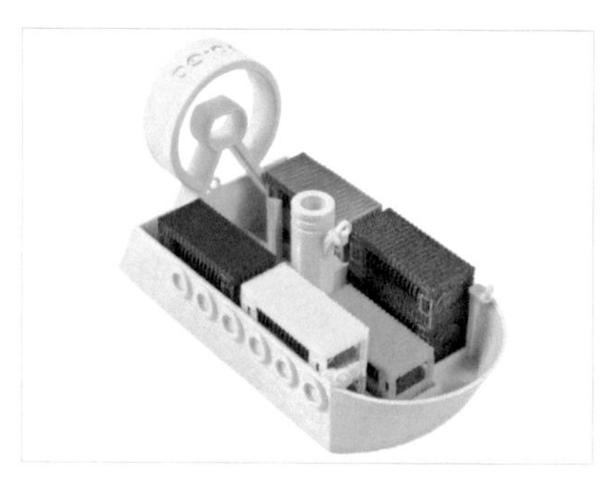

Go-Go AirBoat
Thing: #906745
@Macakcat

Pip Boxes
Thing: #1766512
@Aaskedall

Forest Dragon
Thing: #87458
@DutchMogul

Tide the Fish
Thing: #236345
@Cordavi

Fascinator Hat
Thing: #858954
@Yesquab

THINGIVERSE EDUCATION

Thingiverse Education is a collection of 3D printing projects for the classroom. It's where educators go to find lesson plans, remix them to suit their subjects or grade levels, and share best practices on how to incorporate the technology into their classrooms.

3D printing projects use 3D design and printing in different ways to reinforce core learning objectives in several subjects – from clear STEM applications in engineering and physics, to blended activities in history, music, or foreign languages. Most projects on Thingiverse Education prompt students to plan, design, and print their own solutions to a given problem.

Thingiverse Education organizes projects by grade level, subject, and standards (including Next Generation Science Standards, NGSS and Common Core State Standards, CCSS.) Projects can range in length from a single class period to week-long projects with multiple assignments and objectives.

CREATE A PROJECT ON THINGIVERSE EDUCATION

Whether you design your own 3D model or remix someone else's, every Thingiverse Education project starts with a 3D printable file (.STL). There are some great projects out there, look at existing ones for examples of how to build your own. The projects featured in the second half of this guidebook can all be found on Thingiverse Education.

Here are the steps to create your own Thingiverse Education project for sharing with other teachers.

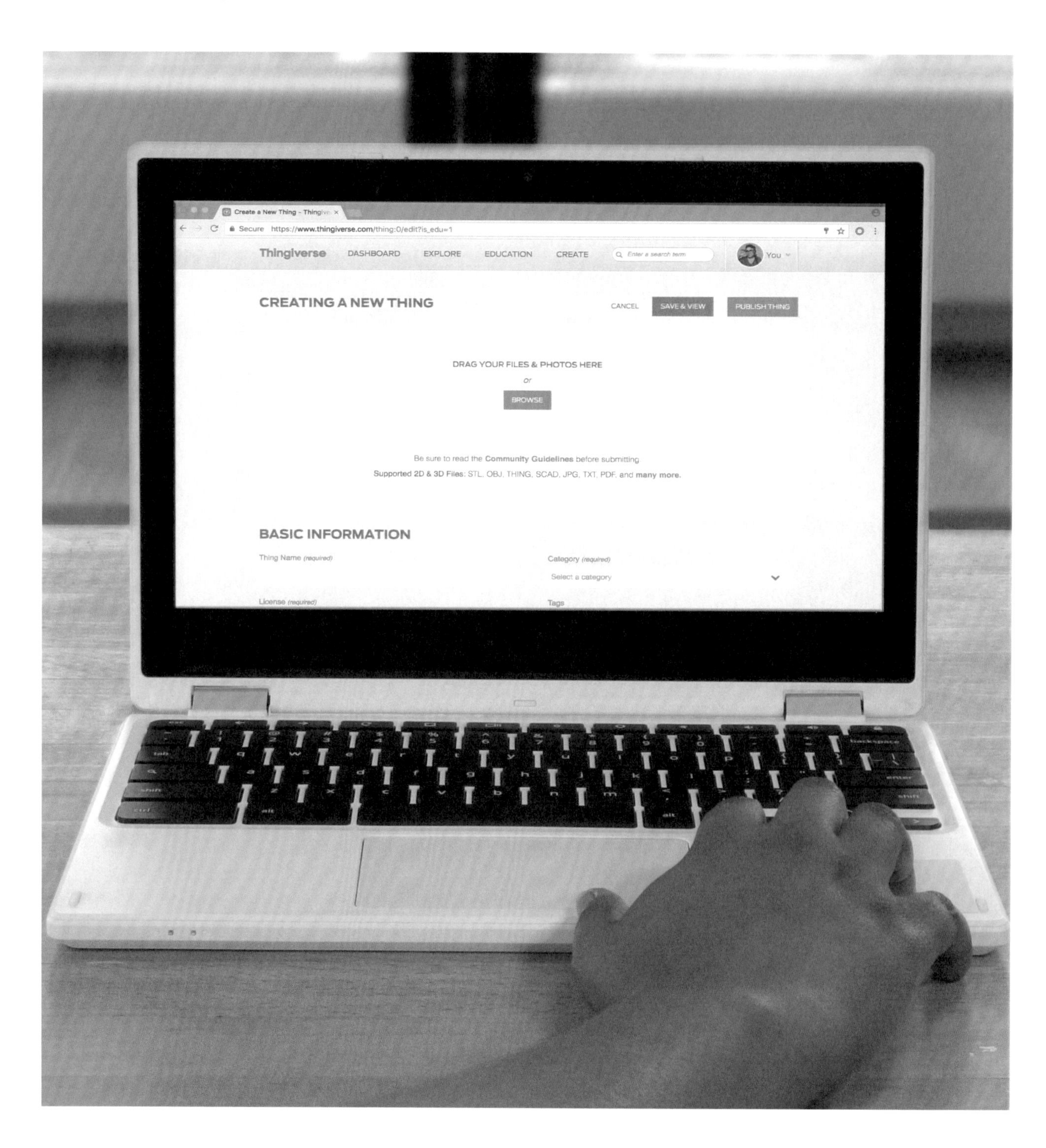

Cloud Types and Display Stands
Designed by @devansic

Grades: 1-4 **Subject:** Science

This project prompts students to design and print models for each of the ten different cloud formations.

SAMPLE PROJECT FROM
Thingiverse.com/Education

01. Create a new Thing:

- Navigate to Thingiverse.com
- Click **Create** then **upload a Thing**

02. Upload your files:

- **.STL file(s):** This could be a starter model, an example of the project's output, a challenge, a tutorial, a demonstration, etc.
- **Photo(s):** This should be a photo of the printed model, or of the project in action. The more photos the better.
- Other files like handouts, rubrics, etc.

03. Complete the Basic Information section:

- Complete the required fields
- Check the box to **submit to Thingiverse Education for approval**

04. Complete the Thing Information sections:

- **Summary:** Short description of the 3D model and the project involved
- **Grade:** Select the recommended grade level for the project
- **Subject:** Include any relevant subjects i.e. science or math
- **Standards:** Select the standards that your project meets
- **Overview and Background:** Give an intro to what students will learn by completing this project
- **Activity / Lesson Plan:** The meat of your project. Be sure to list every step. Remember, a picture is worth 1,000 words
- **Materials Needed:** Itemize the materials for this project

05. Add more information to help other educators:

- **Skills Learned:** List what skills will be gained by completing this project and include specific standards here
- **Duration:** Estimate how long it will take to complete
- **Preparation:** List anything that students should already know or that the teacher needs to do before starting this project
- **Handouts and Assets:** If there are any documents that accompany your project, attach them to the Thing's files and include a brief description here
- **Rubric and Assessment:** Describe what students should have designed, made, printed, or achieved at the end of your project, and any guidelines you have for assessment
- **References:** Include links to valuable resources

06. Publish Thing

3D DESIGN SOFTWARE

Learning how to design in 3D is essential to creating custom 3D printable models. The broad capabilities of the printer will push you to design increasingly complex objects. Once you begin to experiment, you'll discover the unique advantages that different 3D design programs offer for different applications.

We recommend starting with easy solid modeling programs before branching out to digital sculpting or parametric modeling as you become more comfortable. No single program is right for everyone, and it may take a few tries before you find one that you're comfortable with.

There are a lot of 3D design programs available, all with different strengths and weaknesses. When looking at 3D modeling programs, you'll find that all of them fall into three major categories; **solid modeling**, **digital sculpting**, or **polygon modeling**. The projects in this guidebook focus mainly on free programs.

TERMINOLOGY

Solid Modeling: Define and construct solid objects with real world dimensions

Digital Sculpting: Simulate clay sculpting. Push and pull surfaces to create detail and texture

Polygon Modeling: Define outer surfaces like edges and corners to create intricate models

Parametric Modeling: A feature in 3D design programs, use dynamic variables for object parameters so that entire designs can be easily altered or scaled

Mesh: The collection of vertices, edges, and faces that make up the surface of a 3D model

Watertight: A continuous outer surface with no holes in it, necessary for successful 3D printing

Perspective View: Objects further away appear smaller than objects closer to the viewpoint

Orthographic View: Fixes the point of view to a single perspective, where similarly sized objects appear the same size regardless of their distance from the viewpoint

SOLID MODELING

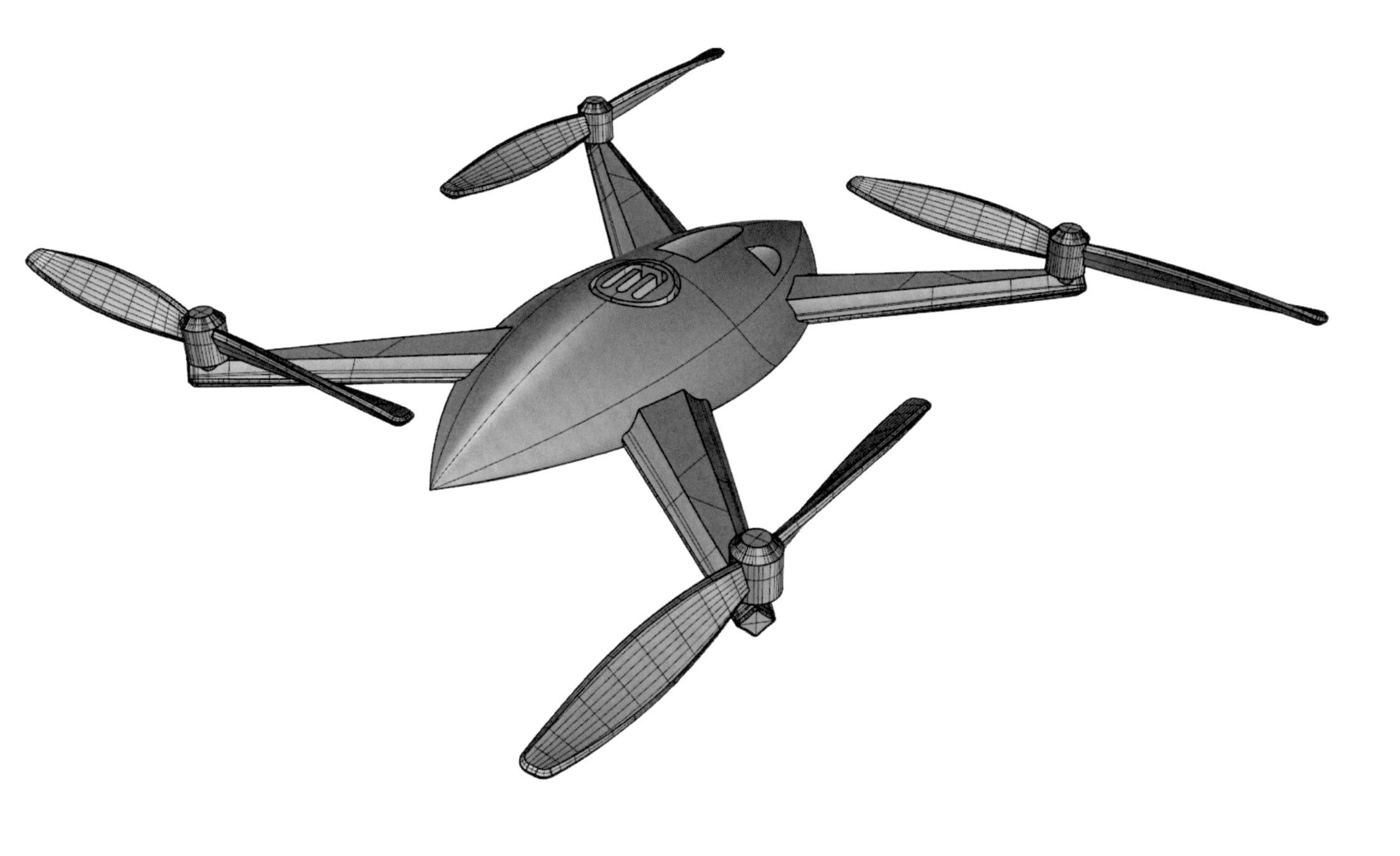

QUADCOPTER DRONE: *Created in Rhinoceros*

Solid modeling programs work well for creating models with real-world dimensions and are used to make functional parts. In some of the advanced programs you can form complex assemblies of objects and run simulations.

Industries: Engineering, industrial design, architecture
Free software: Autodesk Tinkercad™, Autodesk® Fusion 360™, Onshape® and more
Paid software: SolidWorks®, Autodesk Inventor®, Rhinoceros

STRENGTHS

- Creating mechanical structures with dimensions
- Building assemblies
- Simulating real-world conditions
- Access to material property libraries

WEAKNESSES

- Poor organic shape creation
- Difficult to create detailed surface textures and patterns

DIGITAL SCULPTING

ROSE: *Created in Sculptris*

Digital sculpting simulates the process of sculpting with physical clay. Users can push and pull digital clay to create organic, highly detailed and textured models.

Industries: Film, video games, art
Free software: Sculptris™, SculptGL
Paid software: ZBrush®, Mudbox®, 3D-Coat

STRENGTHS

- Highly detailed models
- Organic shapes
- Digital painting

WEAKNESSES

- Creating functional parts is difficult
- Often requires additional hardware like a drawing tablet
- Difficult to design for manufacturing

POLYGON MODELING

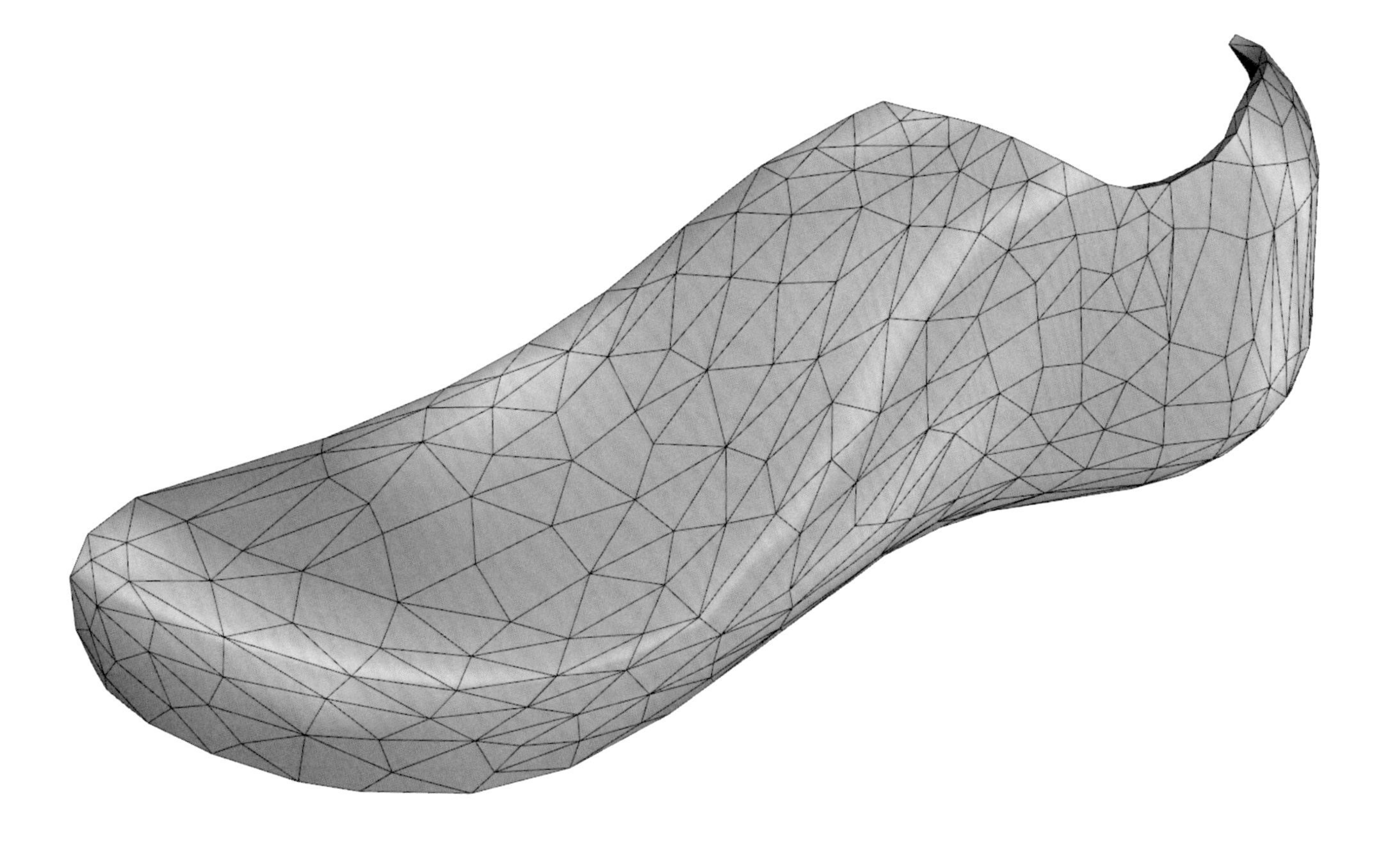

RUNNING SHOE: *Created in Blender*

Polygon modeling gives users direct control of the mesh, faces, vertices, or edges of a model. This allows for the creation of highly detailed and intricate 3D models. These models can be organic or inorganic.

Industries: Animation, visualization, film, video games
Free software: Blender®, Wings 3D
Paid software: Maya®, 3DS Max®, Cinema 4D®

STRENGTHS

› Highly detailed, intricate models

› Direct control of the mesh

WEAKNESSES

› Steep learning curve

› Models aren't always watertight

3D MODELING APPLICATIONS

If you're just getting started, here are some recommended free 3D Design program options to try out.

BLENDER®

Difficulty: Advanced

Type: Polygon Modeling, Digital Sculpting

Platform: Mac, Windows, Linux

Price: Free

Supports modeling, rigging, animation, simulation and even video editing and game creation.

AF

AUTODESK® FUSION 360™

Difficulty: Intermediate

Type: Solid Modeling

Platform: Mac, Windows

Price: Free/$

Powerful software for creating and analyzing complex geometries and assemblies.

AM

AUTODESK® MESHMIXER

Difficulty: Intermediate

Type: Polygon Modeling, Digital Sculpting

Platform: Mac, Windows

Price: Free

Manipulate meshed objects in preparation for 3D printing. Create custom supports, fix mesh errors, and add detail.

MORPHI®

Difficulty: Easy

Type: Solid Modeling

Platform: Mac, iOS, Windows

Price: Free/$

Create 3D designs on mobile devices with this simple and intuitive program.

ONSHAPE®

Difficulty: Advanced

Type: Solid Modeling

Platform: Online

Price: Free/$

Browser-based software for creating and analyzing complex geometries and assemblies.

OPENSCAD

Difficulty: Intermediate

Type: Solid Modeling

Platform: Mac, Windows, Linux

Price: Free

Use code to define object dimensions and dynamic variables for easy resizing and alteration.

SC

SCULPTRIS™

Difficulty: Easy

Type: Digital Sculpting

Platform: Mac, Windows

Price: Free

Push and pull object surfaces. Great for organic, high-detail models. Use SculptGL for a browser-based version.

SKETCHUP®

Difficulty: Easy

Type: Solid Modeling

Platform: Mac, Windows, Online

Price: Free/$

Draw objects with dimensions. Good for accurate, geometric forms, especially architectural models.

TINKERCAD™

Difficulty: Easy

Type: Solid Modeling

Platform: Online

Price: Free

Design in-browser by combining and resizing simple shapes and holes.

MINECRAFT®

Difficulty: Easy

Type: Solid Modeling

Platform: Mac, Windows

Price: $

Though not a traditional 3D modeling program, Minecraft helps to teach core 3D modeling skills.

STUDENT PRIVACY

The projects in Section 2 of this book involve student usage of many 3D design programs. For any program that requires user information and/or account creation, be sure to consult the Terms associated with each program before beginning the project. Students under 13 years of age often are restricted from creating accounts or entering personal information in these programs without explicit consent from their parent/guardian. Here are some suggestions to keep your students' information safe:

DO'S

- Create a generic MakerBot account for students under 13 to use
- Use only school computer and equipment for students under 13

DON'TS

- Allow or require students under 13 to create their own MakerBot account
- Require students under 13 to use their own laptop or mobile devices

PART 02:

3D PRINTING PROJECTS

PROJECT ONE

CLOUD TYPES AND DISPLAY STANDS

LINK: *thingiverse.com/thing:1699444* for access to handouts, videos and other materials associated with this project.

PROJECT INFO

AUTHOR
Danielle Evansic,
@devansic

SUBJECT
Science

AUDIENCE
Grade Levels 1–4

DIFFICULTY
Beginner

SKILLS NEEDED
Basic Tinkercad™ software experience

DURATION
3–4 Class Periods

GROUPS
8 Groups
3–4 Students / Group

MATERIALS
10 No.2 Pencils
Paper and Markers
Index Cards
Camera or Scanner

SOFTWARE
Tinkercad (**web app**)

PRINTERS
Works with all MakerBot® Replicator® 3D printers

PRINT TIME
Prep: 5 hrs / Cloud Base
Lesson: 1–3 hrs / Cloud

FILAMENT USED
1–1 ½ Large Spools

"This project enables kids to do more than design something on the computer. They get to take something they've drawn by hand and turn it into a 3D object. Then, as a bonus, they use their 3D-printed object to show what they know! It's a great, un-scary way of getting kids to make in the classroom."

– **Danielle Evansic**

LESSON SUMMARY

Cloud types differ not only in appearance, but in water content, altitude, and as signals for future weather conditions. Learning about the clouds helps students understand the atmosphere and the interconnectedness between the systems of the planet.

This project prompts students to work in groups to design and print models for each of the ten different cloud formations. The cloud prints are designed to fit on the end of a standard pencil, which can sit in a base that has the name of the cloud formation. The base also has space for flash cards, pictures of clouds, or additional study aids. This is a great science project and can be used to create a display area for teachers who like to use learning stations.

LEARNING OBJECTIVES

After completing this project, students will be able to:

› **Identify** cloud types based on characteristics, appearance, and altitude.

› **Convert** a hand drawing to a 3D model using Tinkercad software.

› **Manipulate** models in Tinkercad using the group, hole, and ruler tools.

NGSS STANDARDS

3-ESS2-1 Earth's Systems Represent data in tables and graphical displays to describe typical weather conditions expected during a particular season.

5-ESS2-1 Earth's Systems Develop a model using an example to describe ways the geosphere, biosphere, hydrosphere, and/or atmosphere interact.

MS-ESS2-4 Earth's Systems Develop a model to describe the cycling of water through Earth's systems driven by energy from the sun and the force of gravity.

3-5-ETS1-1 Engineering Design Define a simple design problem reflecting a need or a want that includes specified criteria for success and constraints on materials, time, or cost.

This project can be extended to meet the following Standards:

5-PS1-1 Matter and Its Interactions Develop a model to describe that matter is made of particles too small to be seen.

5-ESS2-2 Earth's Systems Describe and graph the amounts and percentages of water and fresh water in various reservoirs to provide evidence about the distribution of water on Earth.

MS-ESS2-5 Earth's Systems Collect data to provide evidence for how the motions and complex interactions of air masses results in changes in weather conditions.

MS-LS2-3 Ecosystems: Interactions, Energy, and Dynamics Develop a model to describe the cycling of matter and flow of energy among living and nonliving parts of an ecosystem.

TEACHER PREPARATION

A. Print cloud bases: 3D print each of the cloud bases included in the Thingiverse Education™ post (*thingiverse.com/thing:1699444*). Leave yourself some time for this, as each cloud base will take 4-5 hours to print.

B. Research cloud formations: Have students research and discuss different cloud formations. *NOVA*, *NASA*, and *ABCTeach* offer great resources. You can find these links in the Thingiverse Education post.

C. Make index cards: Have students make index cards for each of the following types of clouds:

› **High-Level Clouds** - Cirrus, Cirrostratus, Cirrocumulus
› **Mid-Level Clouds** - Altostratus, Altocumulus
› **Low-Level Clouds** - Stratus, Stratocumulus
› **Multi-Level/Vertical Clouds** - Cumulus, Cumulonimbus, Nimbostratus

STUDENT ACTIVITY

STEP 01:
DRAW CLOUDS

It's pretty easy to go from a simple drawing to a 3D printed part.

In this project, your group will create a 3D printed model based on your drawing of one cloud type.

A. Choose one of the cloud types (make sure each group selects a different cloud type to work on).

B. Draw and color your clouds. Use dark colors so they show up as solid when designing in Tinkercad software. If you're crafty, leave some blank spots for detail.

TIP: Make sure the cloud names are spaced far enough away from the cloud drawings to allow for easier removal in the Tinkercad software.

STEP 02:
CONVERT DRAWING TO .SVG FILE

A. Take clear, well-lit photos of your drawings.

B. Import the photos to your computer by either emailing them or uploading the photos to a cloud drive like Dropbox®, and save them to your computer.

TIP: Establishing a method to transfer files back and forth between teacher and students (i.e. USB drives, Google Classroom™, etc.) will save you a lot of time.

C. Go to picsvg.com and click **upload picture**.

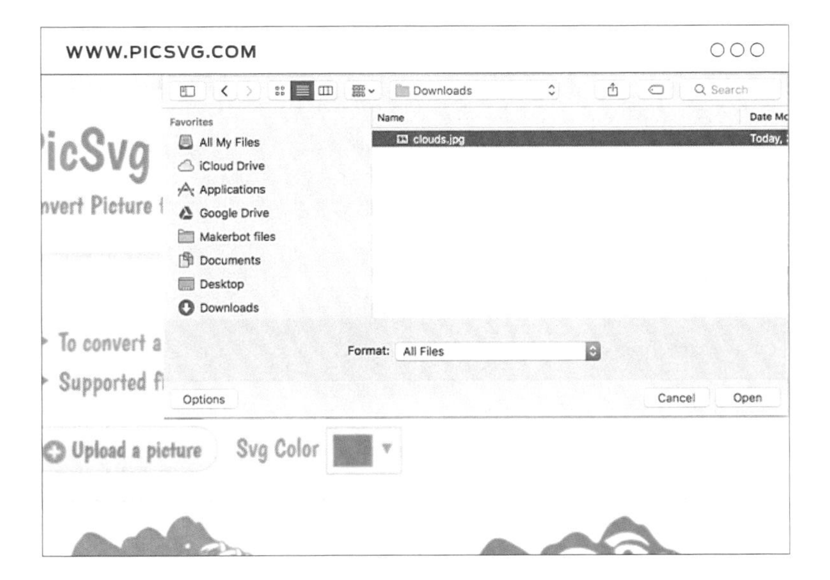

D. Navigate to where the picture is stored on your computer and click **open**.

TIP: Use the filter drop down menu to ensure your cloud is filled in.

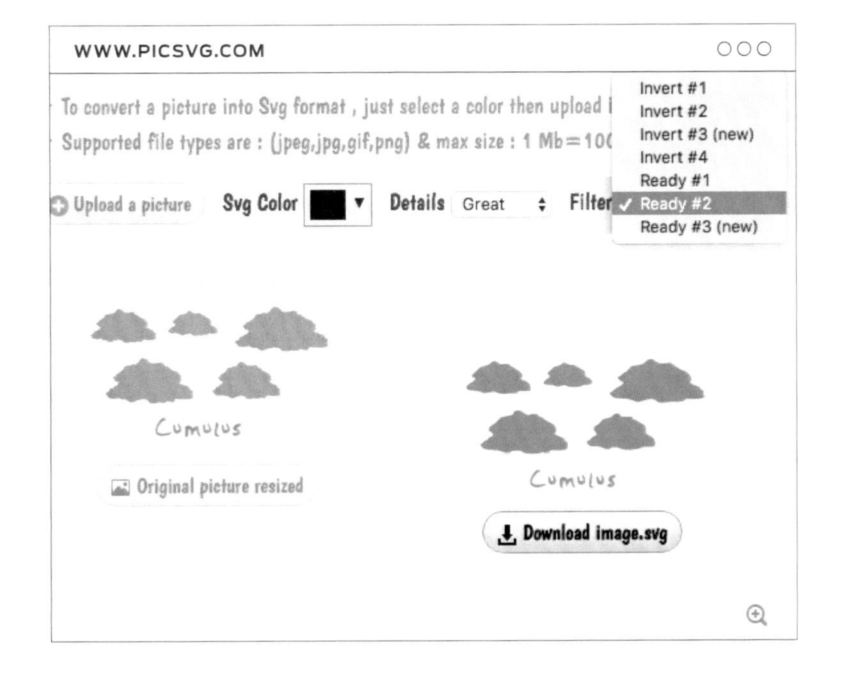

E. Click download image. svg. If the image appears in your browser, right-click the image and select **save as** to save the new .SVG file. Do this for each cloud.

STEP 03:
IMPORT .SVG FILE INTO TINKERCAD

A. Open a browser and navigate to *Tinkercad.com*. Click **create a new design**, and click **import** in the top right of your screen.

B. Select the correct .SVG file, you'll need to choose the scale (size) or dimensions (length, width, height) of the 3D model. Because they can be pretty large, Tinkercad software will give you a suggested size reduction to make your drawing manageable. The cloud files shown in this example were imported at 50% scale.

TIPS: If your import is too big, don't worry, just delete the one you imported, change the numbers, and click **import** again. When re-sizing, be mindful of your build plate size.

When your drawing is imported, only dark colored parts will be visible. If they aren't connected, they will be separate parts when printed. We will have to put in connectors later so that they stay together when printed.

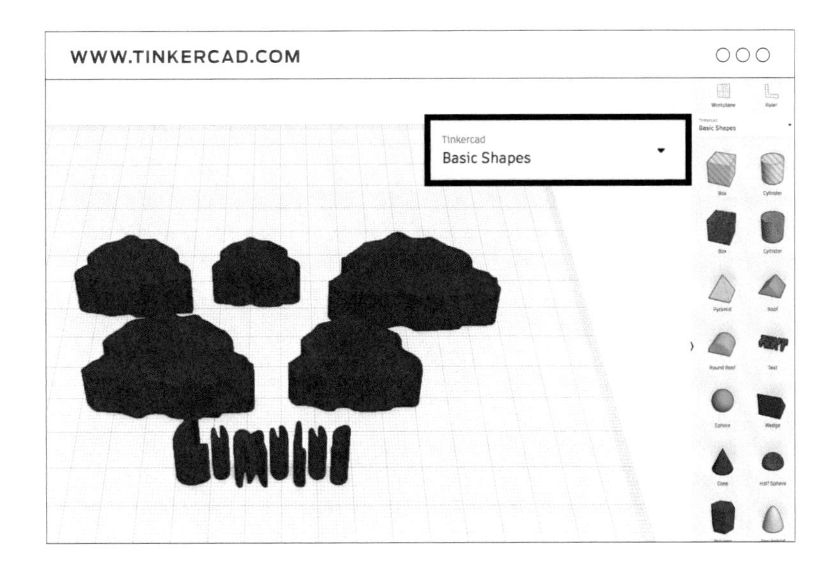

C. Remove anything extra that came in from the drawing (names, extra lines, etc.) To do this, go to the **basic shapes** menu, select a **box** and make sure it's big enough to cover.

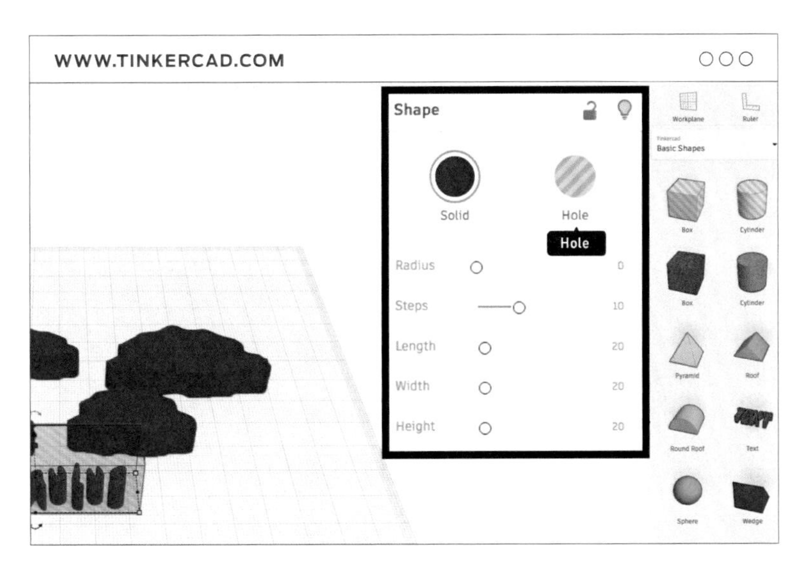

D. Turn the box into a hole by selecting the box, then clicking **hole** in the dialogue window near the top right corner.

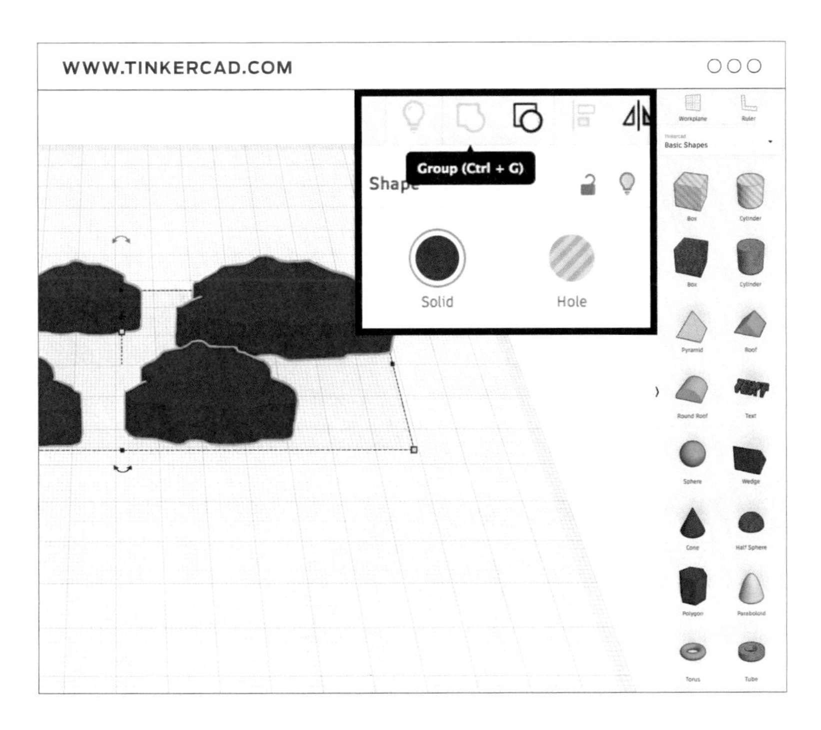

E. Select the cloud and hole by holding **shift** while clicking on both, and use **group** to combine them. This will remove both the box and everything it covers.

STEP 04:
ADD CLOUD CONNECTORS

Now that all that clutter is hidden, add some connectors to the clouds. You might be able to skip this step if your cloud is already one single part.

A. Drag a new box onto your workplane and position it between two cloud pieces.

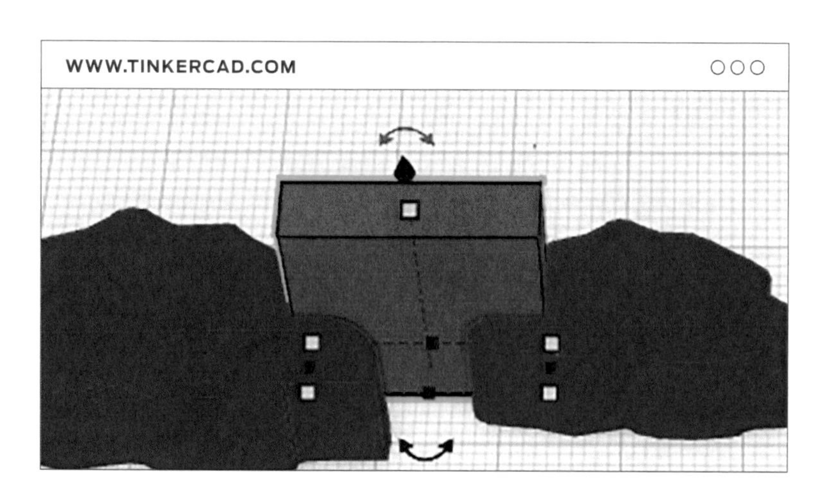

TIP: Use the **right click** on your mouse to orbit your view to make sure everything is in the correct place.

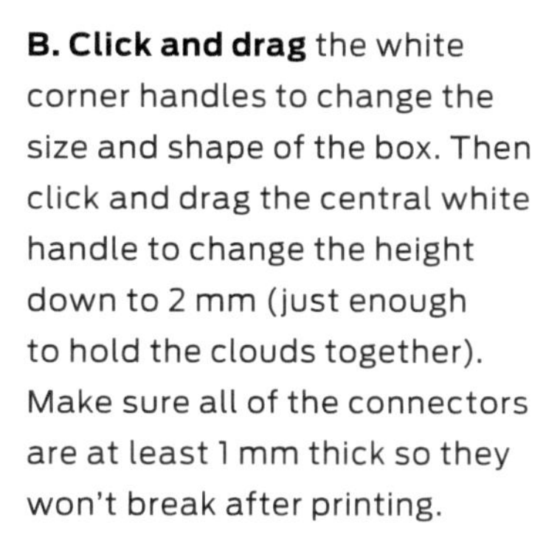

B. Click and drag the white corner handles to change the size and shape of the box. Then click and drag the central white handle to change the height down to 2 mm (just enough to hold the clouds together). Make sure all of the connectors are at least 1 mm thick so they won't break after printing.

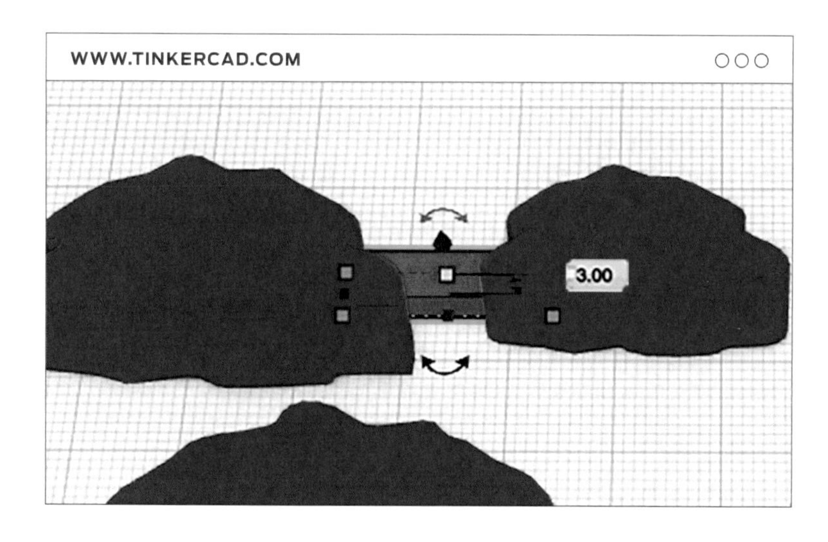

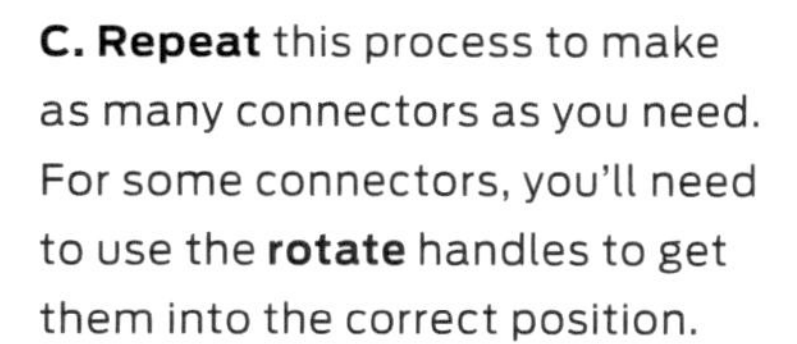

C. Repeat this process to make as many connectors as you need. For some connectors, you'll need to use the **rotate** handles to get them into the correct position.

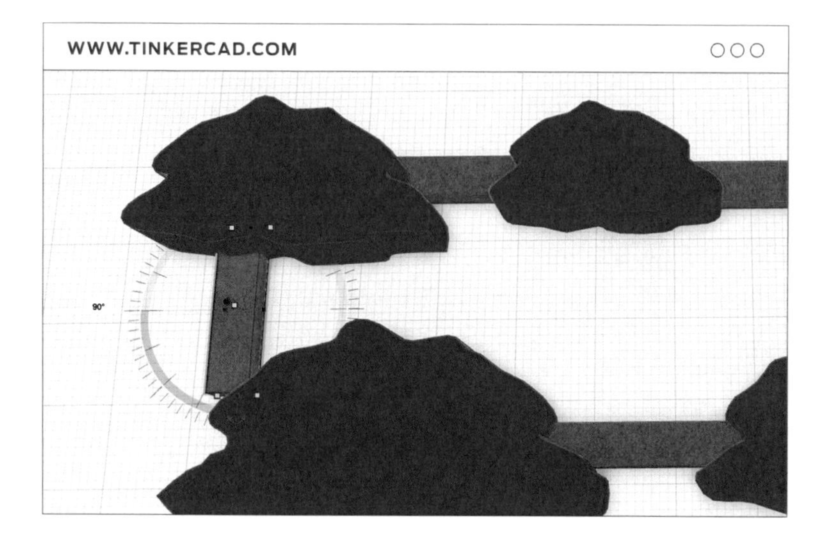

TIP: Make sure all connectors lay flat on the workplane and are not floating in the air. They should share the same bottom surface as the clouds. Use the hotkey **D** to drop a selected item to the workplane.

STEP 05:
MAKE PENCIL TOPPER

Most pencils are a little less than 8 mm in diameter. Add a couple of cylinders to make the topper for the pencil.

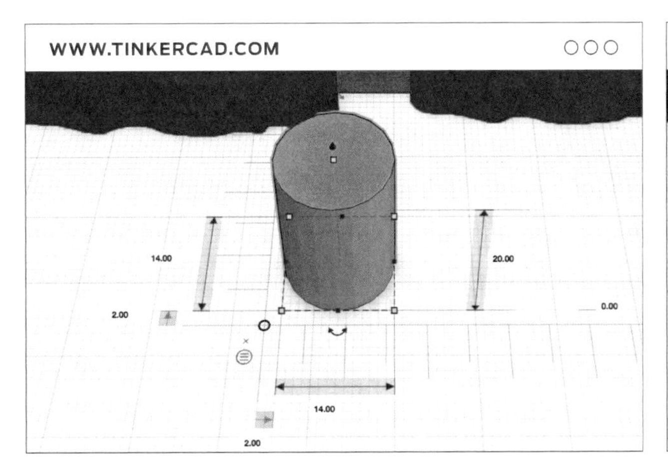

A. Drag a new box over from the **basic shapes** menu, then drag a **ruler** to help you out. Click on the cylinder to apply the ruler to it. Change the diameter of the cylinder to 14 mm.

TIP: With the ruler, you don't need to drag the corners of your object to resize. You can just click on the dimensions you want to change and type in the measurements you want.

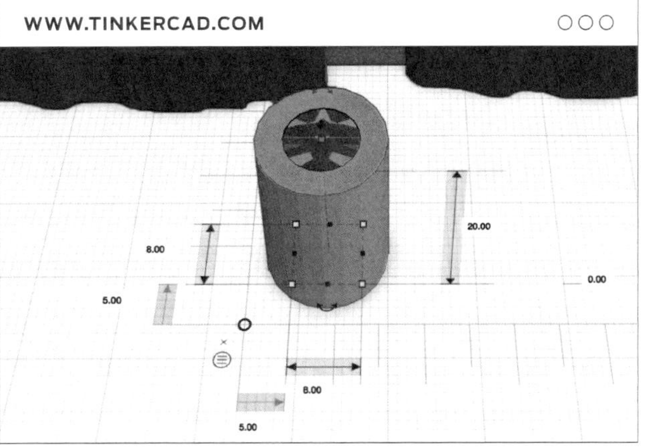

B. Make a cylinder hole for the pencil to fit into. Select the first cylinder and duplicate it by pressing **ctrl-C then ctrl-V (cmd-C, cmd-V on Mac)**. Change the size of the second cylinder to 8 mm in diameter and select hole. This will be the cutout for the pencil to fit into.

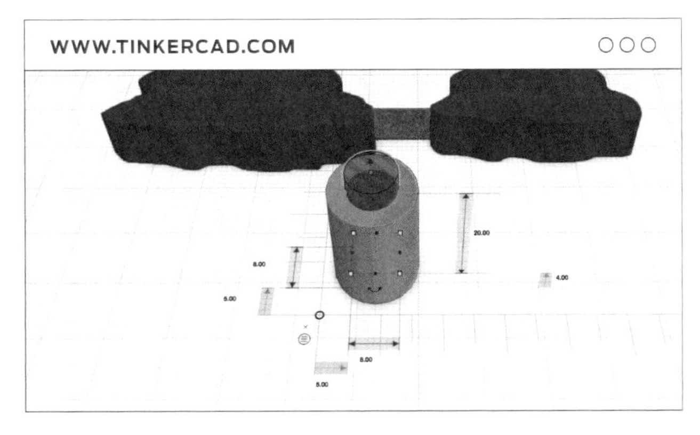

C. Change the elevation of the hole so that it sits 3 mm above the workplane. You do this by clicking on the black cone at the top center of the cylinder and dragging it up 3 mm.

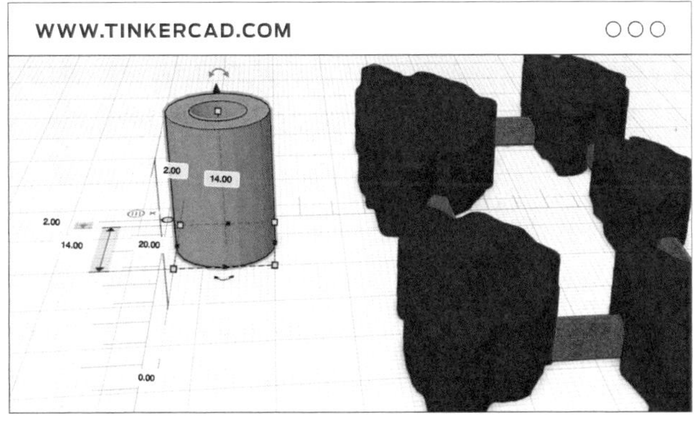

D. Group the cylinders once you're done moving and resizing them. To do this, select both cylinders and click on the **group** icon at the top of the page.

STEP 06:
PRINT

Now it's time to export the model so you can print it.

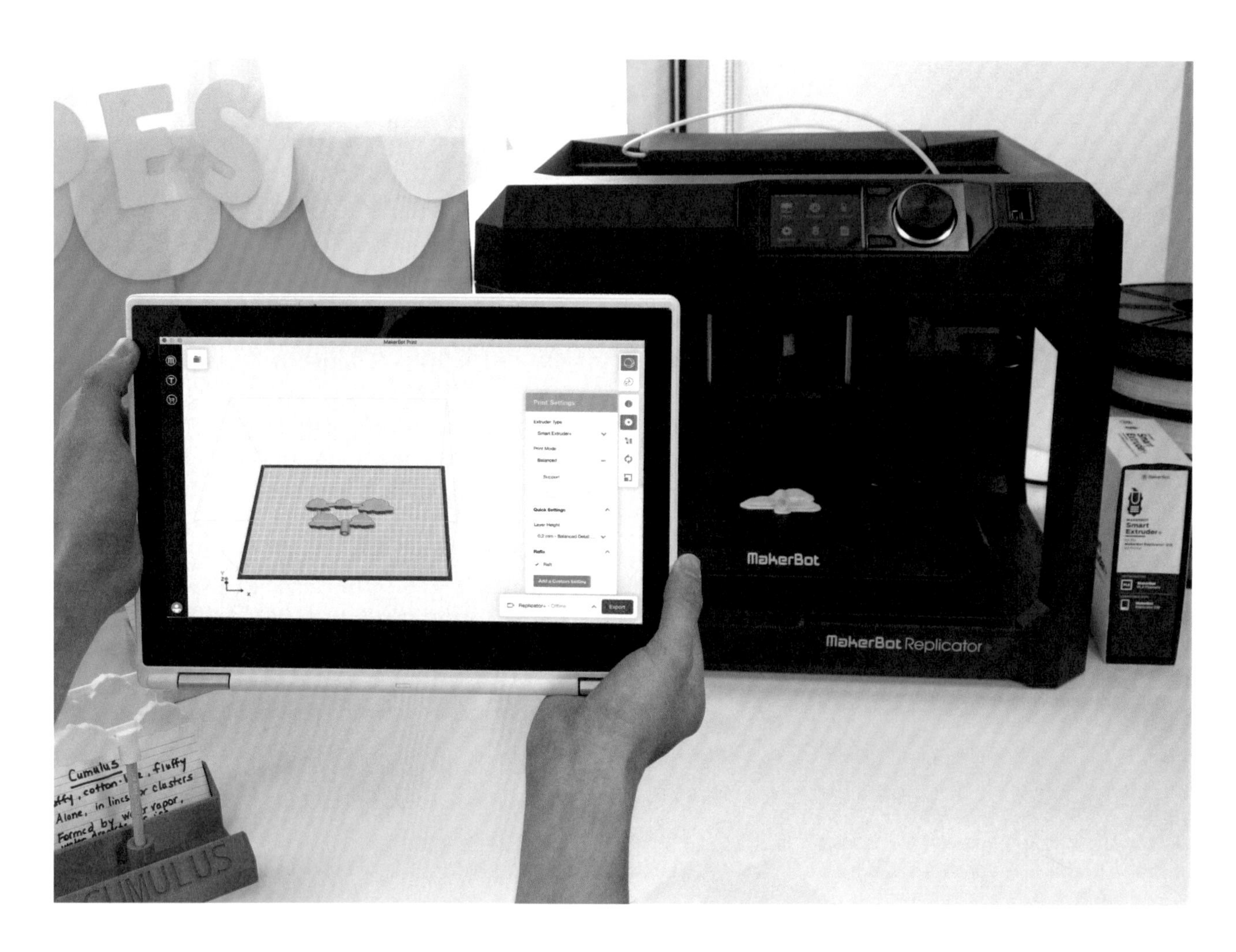

A. Click export in the upper right side of the Tinkercad software window. Select the format that your printer uses. (MakerBot Print imports .STL files.) The download will start once you click on the desired file format.

B. Import the files into MakerBot Print™, select your print settings, and start printing.

Print Settings:

Rafts	Yes
Supports	No
Resolution	0.2 mm
Infill	10%

TIPS: The export window in the Tinkercad software window doesn't disappear until you close it, so be sure to check your downloads folder before you click export again (or you may end up with some extra copies).

You can also name your file in the Tinkercad software, unless you like those way-cool file names the program randomly assigns you. While they are entertaining, they're not always descriptive enough.

PROJECT COMPLETE:
REACH FOR THE SKY!

Once printed, assemble the clouds onto the pencils and put them onto their respective bases. If you want, you can cut the pencils to different lengths to match the altitudes of the clouds (low, medium, high). Nice job! You're ready to go!

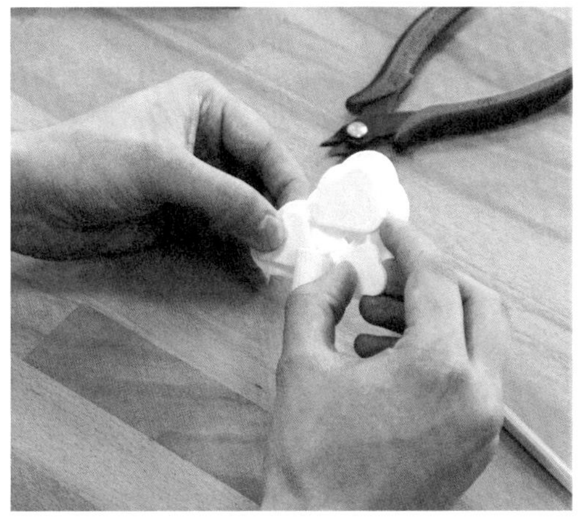

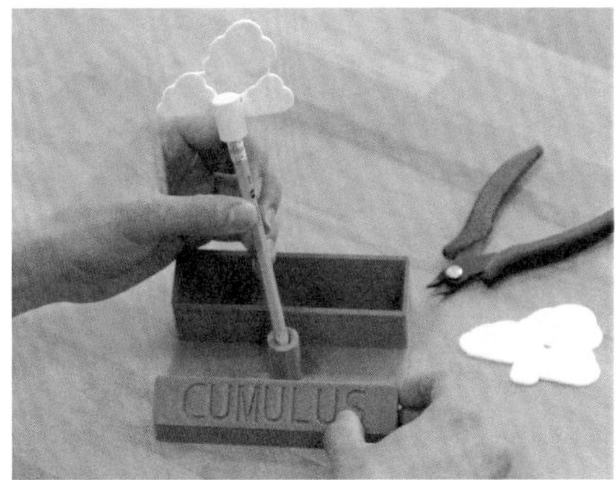

GOING FURTHER

A. After the project, the cloud models and bases can be used as a learning station that students will cycle through over a series of days. As they start the activities, students can be given puzzles and worksheets that encourage them to explore the models on their own, gathering information while they complete their assignments.

B. Consider scaffolding activities, where students can match clouds with the bases where the pencils are attached, or sort the note cards into the bases with the clouds. Additionally, a summative activity could include having the models, pencils, and bases disassembled and asking the students to work together to place the components in the correct positions.

C. Assess students with quizzes or performance tasks based on recall of the cloud names and characteristics. Or have them perform a puppet show with the clouds, where the characters in the show talk about what clouds they see and how those reflect the weather or anticipated weather changes.

PROJECT TWO

3D MUSICAL SHAPES

LINK: *thingiverse.com/thing:1753157* for access to handouts, videos and other materials associated with this project.

PROJECT INFO

AUTHOR
Karie Huttner,
@*khuttner*

SUBJECT
Music, Math, Art

AUDIENCE
Grade Levels K–5

DIFFICULTY
Beginner

SKILLS NEEDED
None

DURATION
1-2 Class Periods

GROUPS
15 Groups
1–2 Students / Group

MATERIALS
iPad®
Translucent Filament
(**recommended**)
Seed Beads

SOFTWARE
Morphi® for iPad (**iOS**)
Tinkercad (**web app**)

PRINTERS
Works with all
MakerBot® Replicator®
3D printers

PRINT TIME
Approximately 1.5 hrs
per instrument

FILAMENT USED
½ Large Spool

"I love doing this project with my students because not only can you create a musical shape, but it also gives students the opportunity to learn about infill and the 3D printing process."

– **Karie Huttner**

LESSON SUMMARY

This lesson is an introduction to 3D shapes and creating music using the structure of a 3D print. Students will select a 3D shape to print, then pause the print and add beads to create a musical instrument.

LEARNING OBJECTIVES

After completing this project, students will be able to:

› **Understand** and create 3D shapes using Morphi.

› **Resize** 3D models using the ruler tool.

› **Learn** about 3D printing and print pausing.

COMMON CORE MATH STANDARDS

K.G.A.3 Geometry Identify shapes as two-dimensional (lying in a plane, "flat") or three-dimensional ("solid").

K.G.B.4 Geometry Analyze and compare two- and three-dimensional shapes, in different sizes and orientations, using informal language to describe their similarities, differences, parts (e.g., number of sides and vertices/"corners") and other attributes (e.g., having sides of equal length).

K.MD.A.1 Measurement & Data Describe measurable attributes of objects, such as length or weight. Describe several measurable attributes of a single object.

1.G.A.2 Geometry Compose two-dimensional shapes (rectangles, squares, trapezoids, triangles, half-circles, and quarter-circles) or three-dimensional shapes (cubes, right rectangular prisms, right circular cones, and right circular cylinders) to create a composite shape, and compose new shapes from the composite shape.

2.MD.A.1 Measurement & Data Measure the length of an object by selecting and using appropriate tools such as rulers, yardsticks, meter sticks, and measuring tapes.

TEACHER PREPARATION

A. Install the Morphi app on student iPad(s).

TIP: If students don't have iPads, you could use the Morphi app for Windows or Mac.

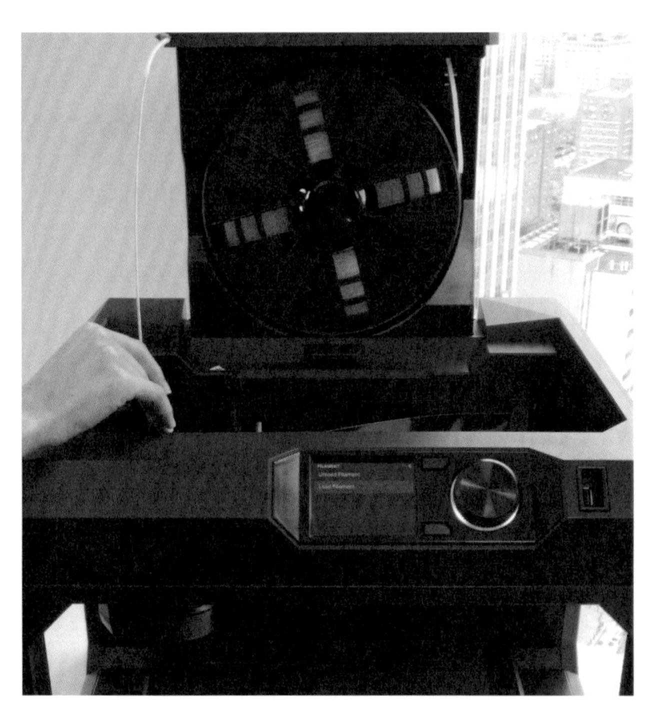

B. Load your 3D printers with filament, translucent if you have it.

C. Lay out beads for students to choose.

STUDENT ACTIVITY

STEP 01:
CREATE WITH THE MORPHI APP

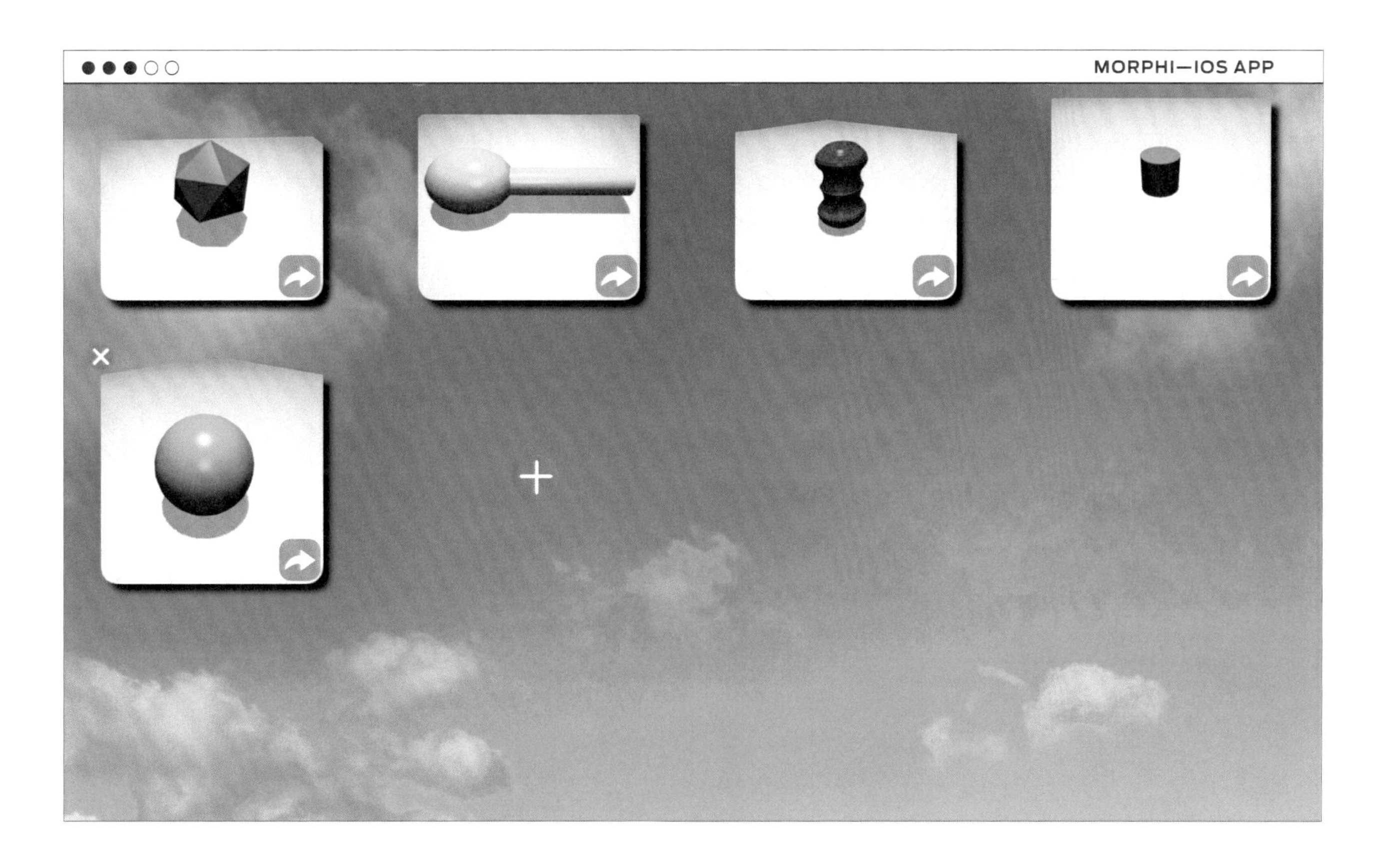

A. Create new design. Tap on the plus symbol to begin a new design.

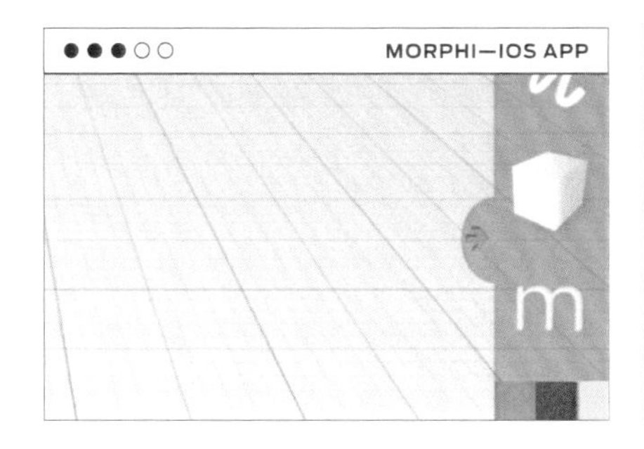

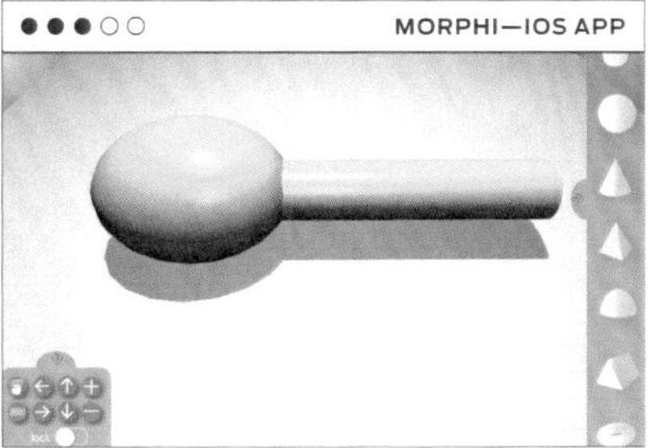

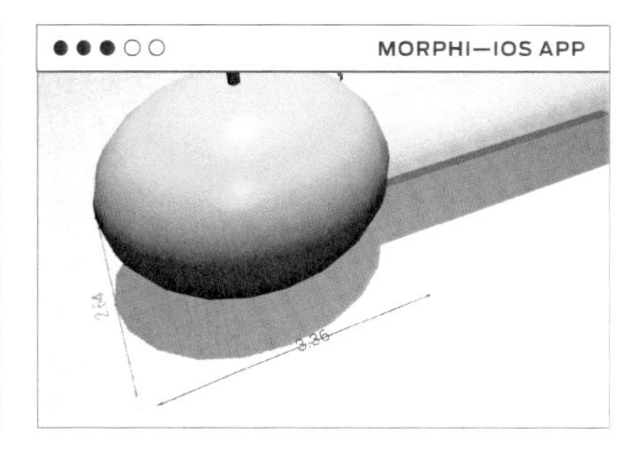

B. Select shape: Select the shapes menu by clicking the cube on the right. Choose one shape that you want to create from the shape choices and tap to place it on the workplane.

C. Practice: Spend some time getting used to the navigation panel on the lower left corner. You may need to delete and re-add objects as you are learning.

D. Adjust Size: Tap on your shape to select it. Tap on the **ruler tool** in the top panel and select inch. Then resize the shape to roughly 2 inches tall. Practice with the slider to adjust the size of your shape.

STEP 02:
EXPORT THE DESIGN

You will need to export the design for 3D printing from Morphi as an .STL file and share it with your teacher for final printing.

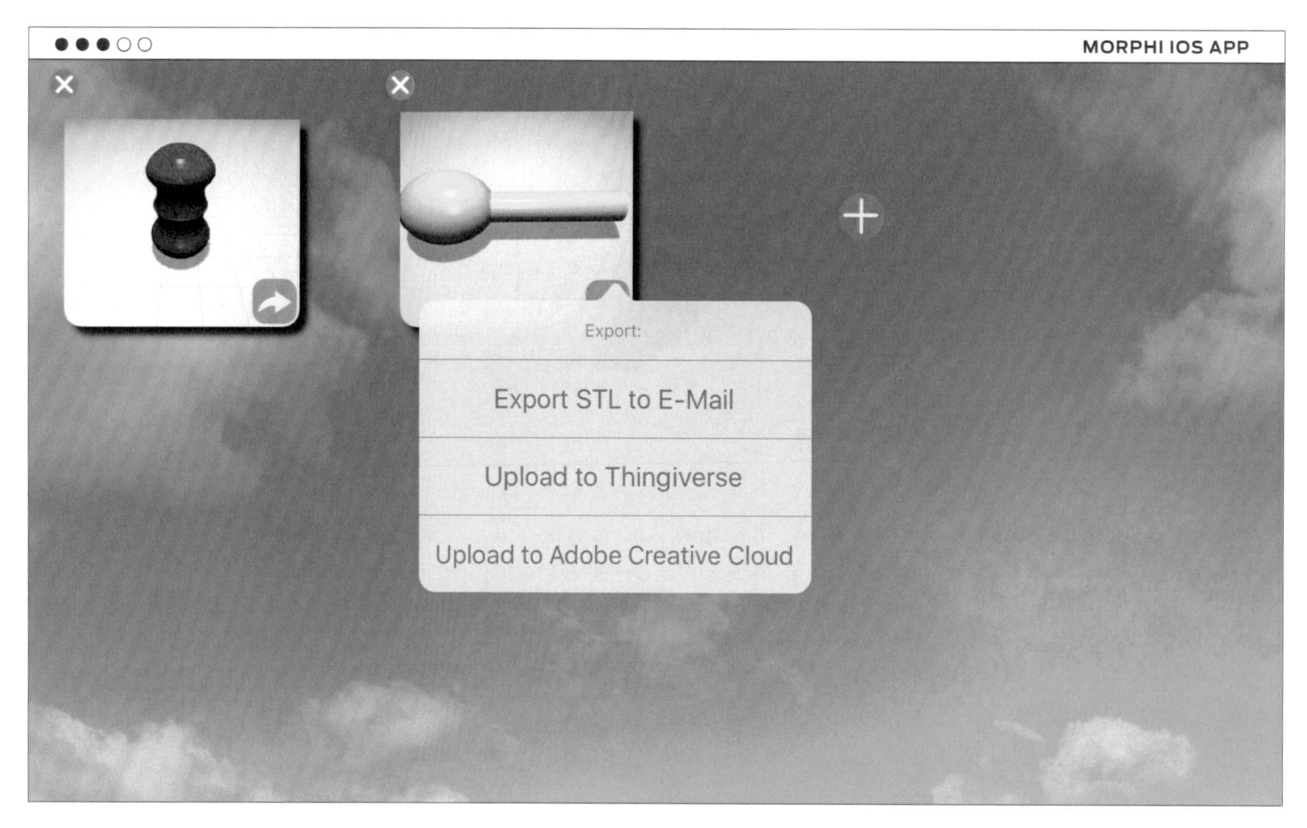

A. Save file: Tap the save button at the top of the screen to get back to the home screen.

B. Export file: Tap on the small arrow on the bottom right of the project icon. Choose to email the .STL file directly to your teacher or upload to Thingiverse.

STEP 03:
PREPARE THE DESIGN IN MAKERBOT PRINT

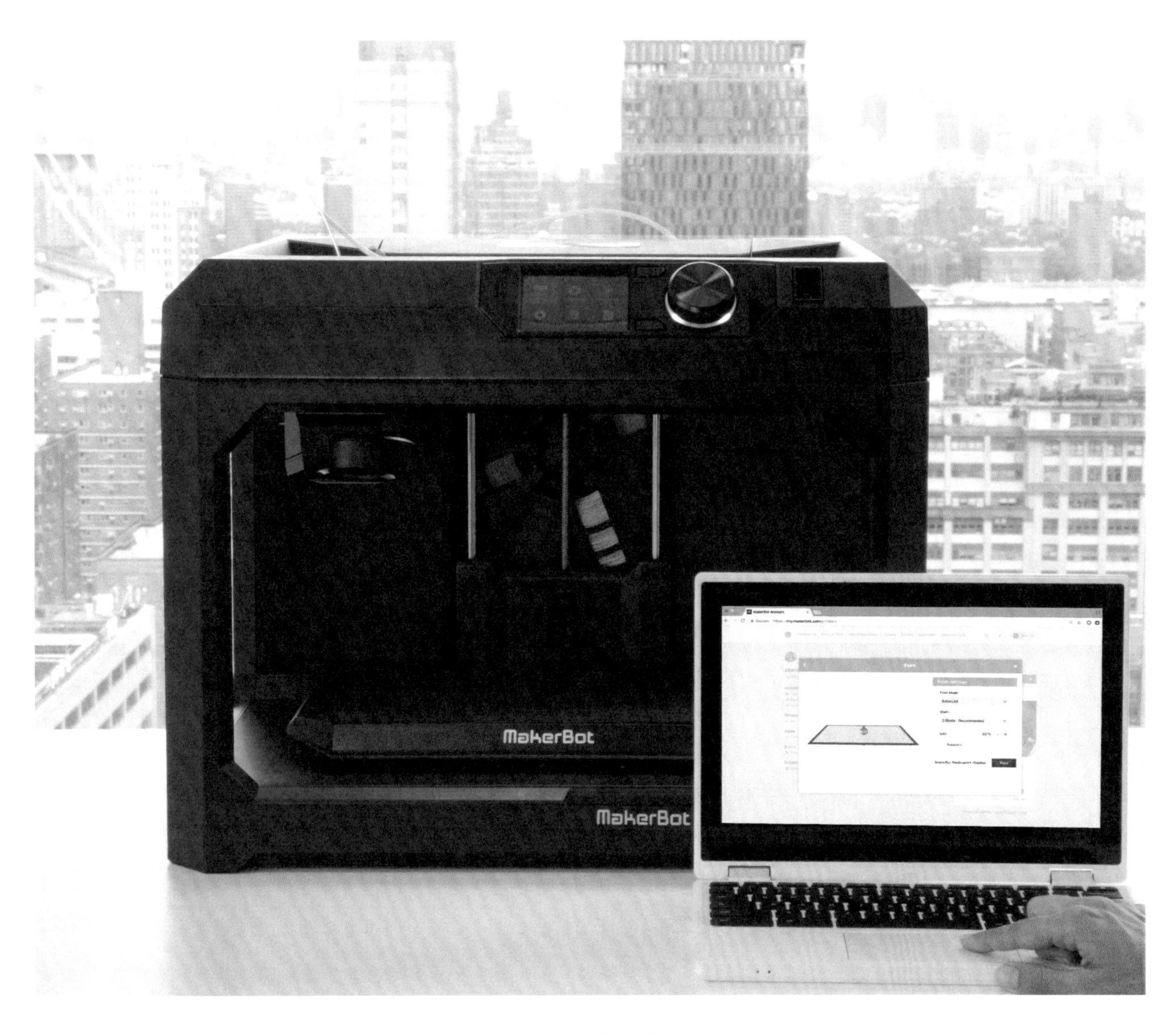

A. Import into MakerBot Print to prepare for printing.

TIP: Rotate the part to make sure there is a flat surface on the bottom for printing. Use the **place face** on build plate tool in MakerBot Print to help. Review **print preview** before printing to understand what infill will look like when paused.

B. Adjust print settings: Decide what settings you'd like for your instrument. Lower infill will allow more space for beads.

Print Settings:

Rafts	Yes
Supports	Depends on Model. For spheres, supports are recommended.
Resolution	0.2 mm
Infill	10%

STEP 04:
PRINT

A. Start your print: Printing will take approximately 1–1.5 hours per model depending on the shape.

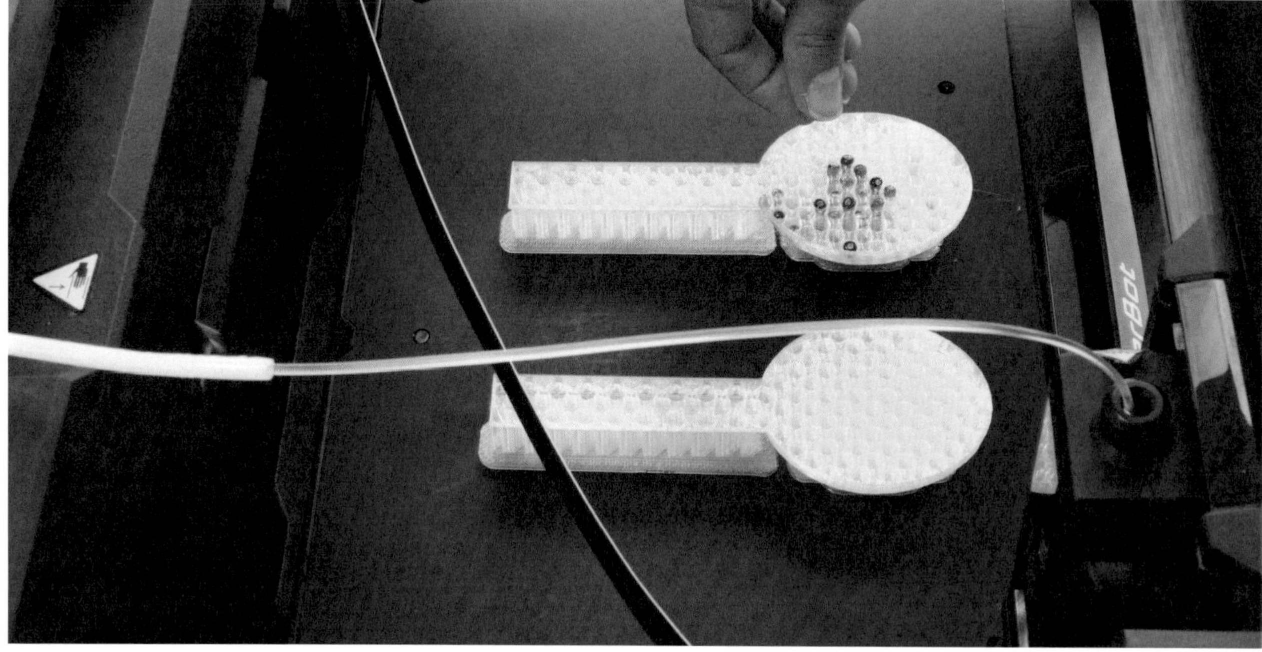

B. Pause print: As the models are being printed, pause the print about halfway through to add beads. After adding the beads, resume the build.

CAUTION: When the printer is paused, the extruder is still very hot. Carefully insert the beads through the top or side of the printer. Make sure the beads are not overflowing, otherwise they will knock into the extruder when the print resumes. The teacher should handle this part.

PROJECT COMPLETE:
SHAKE IT UP!

GOING FURTHER

A. Create more advanced shapes in Morphi and Tinkercad.

B. Use the draw feature in Morphi to hand-draw shapes to be used as instruments.

C. Compare the sounds made by the different printed objects based on different variables:

› Same shape, different sizes

› Same size, same shape, different amount of beads

› Same size, same shape, different infill density or pattern

PROJECT THREE

GO FROM CODE TO CAD

LINK: *thingiverse.com/thing:2072299* for access to handouts, videos and other materials associated with this project.

PROJECT INFO

AUTHOR
Ryan Erickson
@*learningblade*

SUBJECT
Technology, Math, Geometry

AUDIENCE
Grade Levels 3–9

DIFFICULTY
Beginner

SKILLS NEEDED
None

DURATION
1–2 Class Periods

GROUPS
15 Groups
1–2 Students / Group

MATERIALS
Sketchbook
Ruler

SOFTWARE
Turtle Blocks
Tinkercad
(**web apps**)

PRINTERS
Works with all MakerBot® Replicator® 3D printers

PRINT TIME
Prep: 0 hrs
Lesson: 1 hr / Shape

FILAMENT USED
½ Large Spool

"Coding has become a valuable skill both for its growing use in STEM jobs and for teaching computational thinking skills. This Code to CAD project allows my students to hold, interact with, and feel their coding skills, which brings a whole new level of interaction to their learning that they could never have without the use of a 3D printer."

– **Ryan Erickson**

LESSON SUMMARY

Learning the basic logic that governs code can be easy and visually stimulating. In this project, you'll use an in-browser, block-based coding program called Turtle Blocks to create a geometric shape. Without needing to learn a specific coding language, you can intuitively string together command blocks that instruct a cursor (in this case, a turtle) to draw the object you want on the screen.

By asking students to create a shape that meets certain requirements, this part of the project is easily scalable from a simple perimeter demonstration to a more advanced proof of triangle relationships.

After completing the coding and geometry task, students can export their shape and import it into Tinkercad software to easily create a 3D printable file. Being able to hold an object you created using code, then CAD, is an especially powerful moment—one that will inspire students to practice computational thinking and dramatically advance their understanding of technology.

LEARNING OBJECTIVES

After completing this project, students will be able to:

› **Code with Turtle Blocks** to create a single shape, figure, or pattern

› **Use their knowledge** of geometry to create complex shapes with code

› **Manipulate** that shape in Tinkercad software and prepare it for 3D printing

NGSS STANDARDS

3-5-ETS1-3: Engineering and Design Plan and carry out fair tests in which variables are controlled and failure points are considered to identify aspects of a model or prototype that can be improved.

COMMON CORE STANDARDS

4.MD.2: Measurement and Data Use the four operations to solve word problems involving distances, intervals of time, liquid volumes, masses of objects, and money, including problems involving simple fractions or decimals, and problems that require expressing measurements given in a larger unit in terms of a smaller unit.

4.MD.3 Measurement and Data Apply the area and perimeter formulas for rectangles in real world and mathematical problems.

4.G.A.1 Geometry Draw points, lines, line segments, rays, angles (right, acute, obtuse), and perpendicular and parallel lines. Identify these in two-dimensional figures.

TEACHER PREPARATION

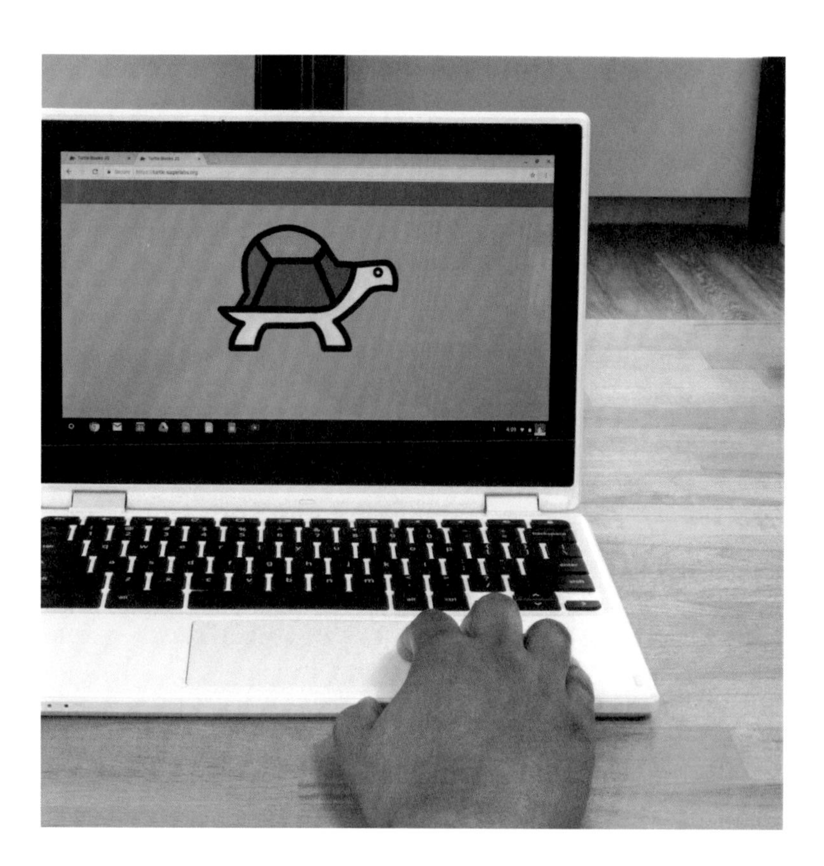

A. Open a browser and go to turtle.sugarlabs.org

B. Explore the different functions for moving and drawing. For an extensive tutorial and activities for Turtle Blocks, or for links to similar programs, check out the Thingiverse Education post (thingiverse.com/thing:2072299).

C. Experiment until you've learned to prepare a shape that validates principles of perimeter, fractions, spirographs, trigonometry, vectors, or coordinate geometry.

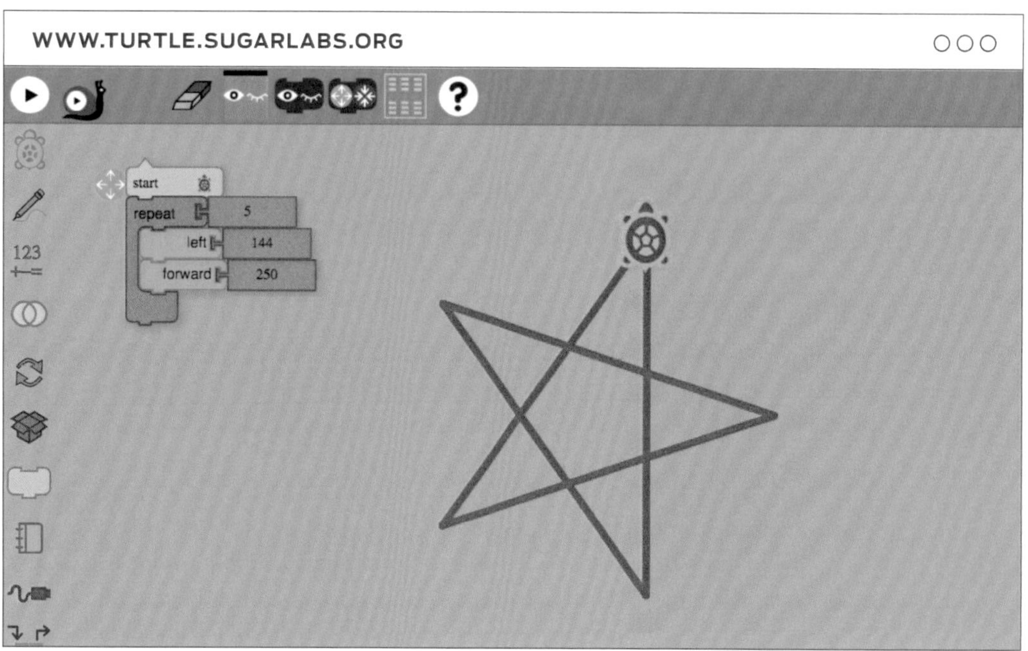

D. Run a demonstration of Turtle Blocks for your students. Prompt them with these questions to guide their understanding: How does the turtle know what to do on the screen?

› How does block coding work?

› What can you design using the blocks provided?

E. Create a Tinkercad account

STUDENT ACTIVITY

STEP 01: PICK A SHAPE

A. Pick a geometric shape to create with code. Choose a simple one to start with; you can try for more advanced shapes later on.

B. Open Turtle Blocks and begin experimenting with the different commands under the **turtle**, **pen**, and **flow** menu options.

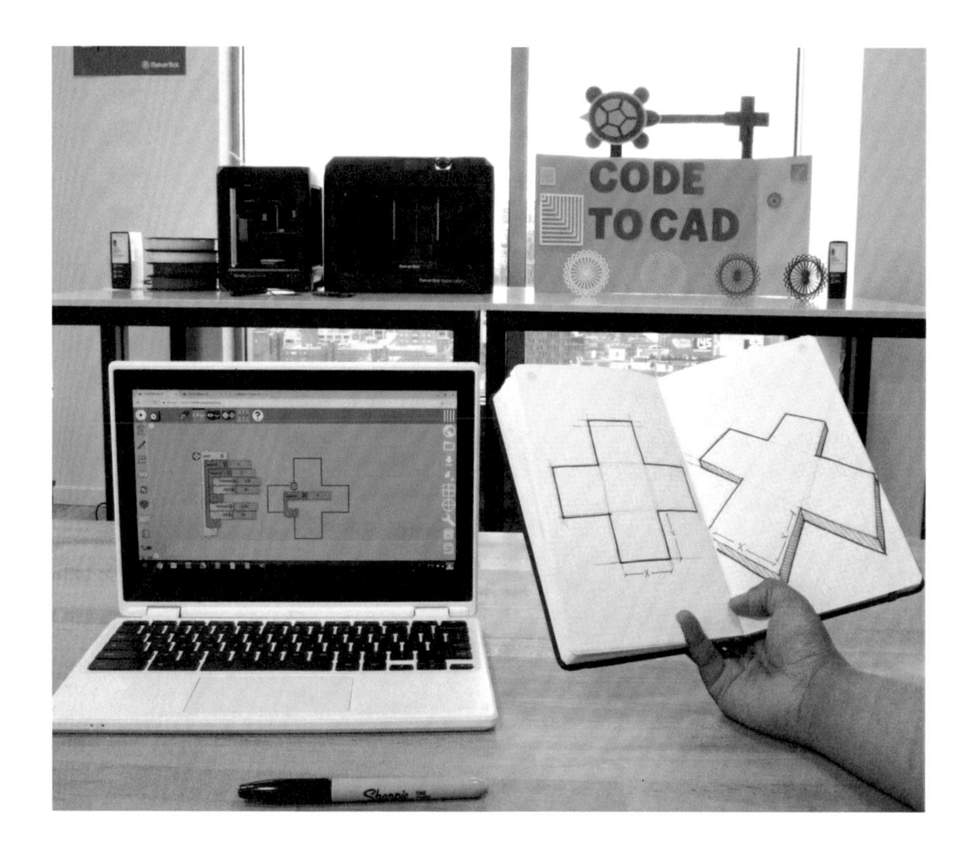

C. Experiment with the following basic functions:

› **forward / back**
› **left / right**
› **pen up / down**
› **repeat**

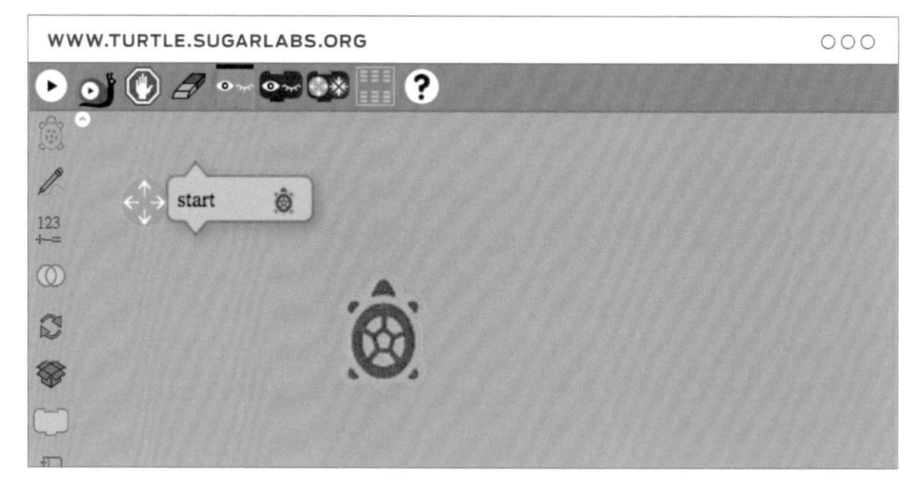

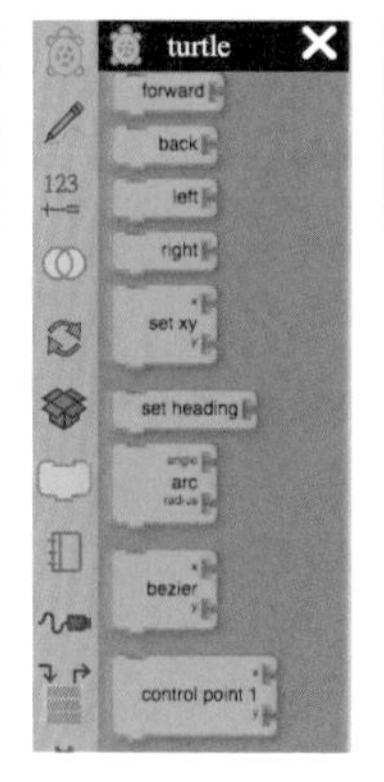

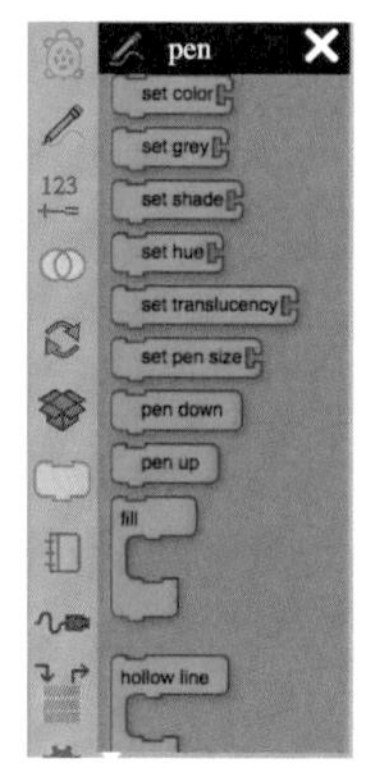

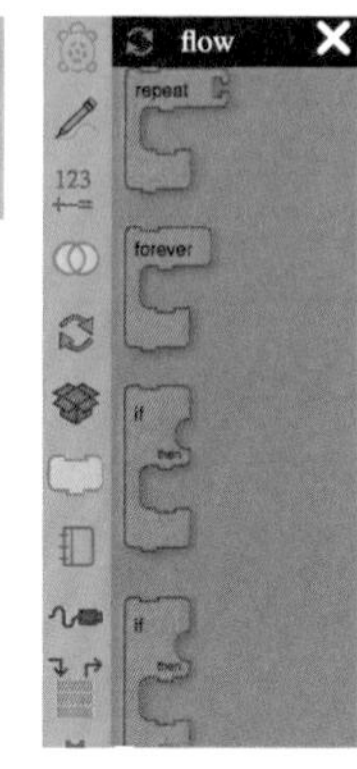

STEP 02:
CODE YOUR SHAPE

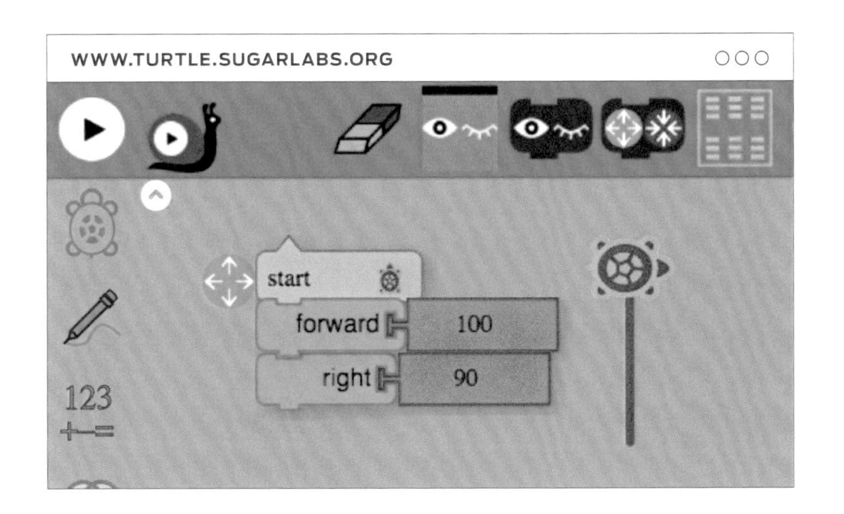

A. Drag commands like **forward** and **right** onto the screen, then connect them under the yellow **start block** to create an operation.

TIP: Use the **eraser** to clear your canvas if you need to restart.

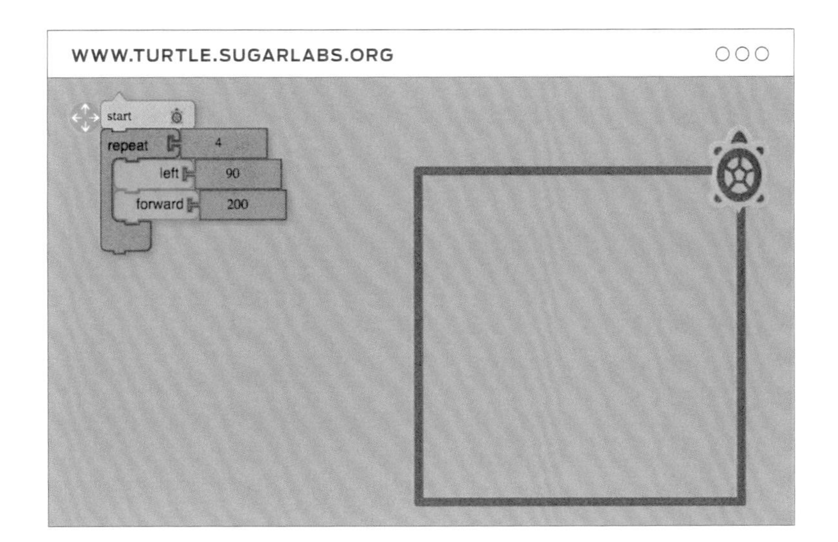

B. Create simple objects by using this drawing function until you've created your desired shape.

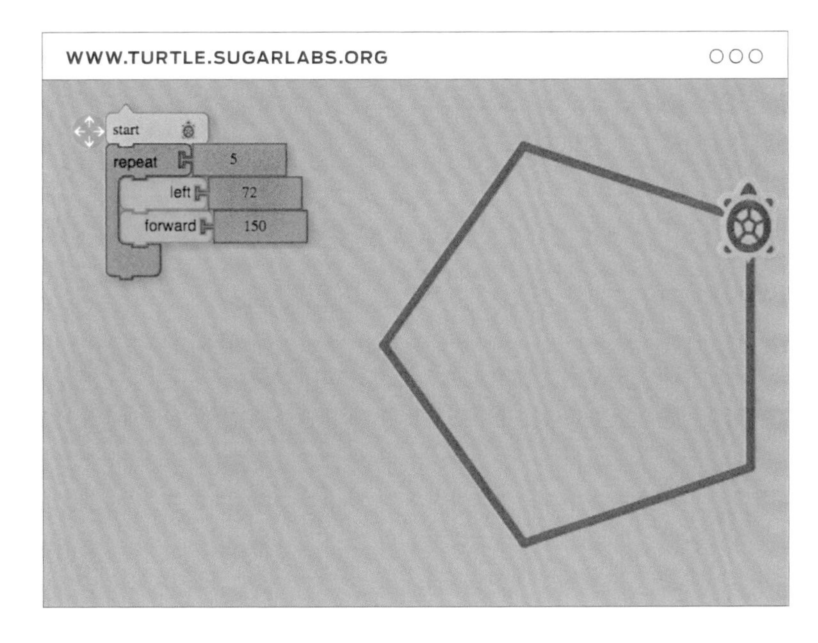

C. Practice making shapes requiring multiple different moves. Try using the **repeat** command to simplify your code.

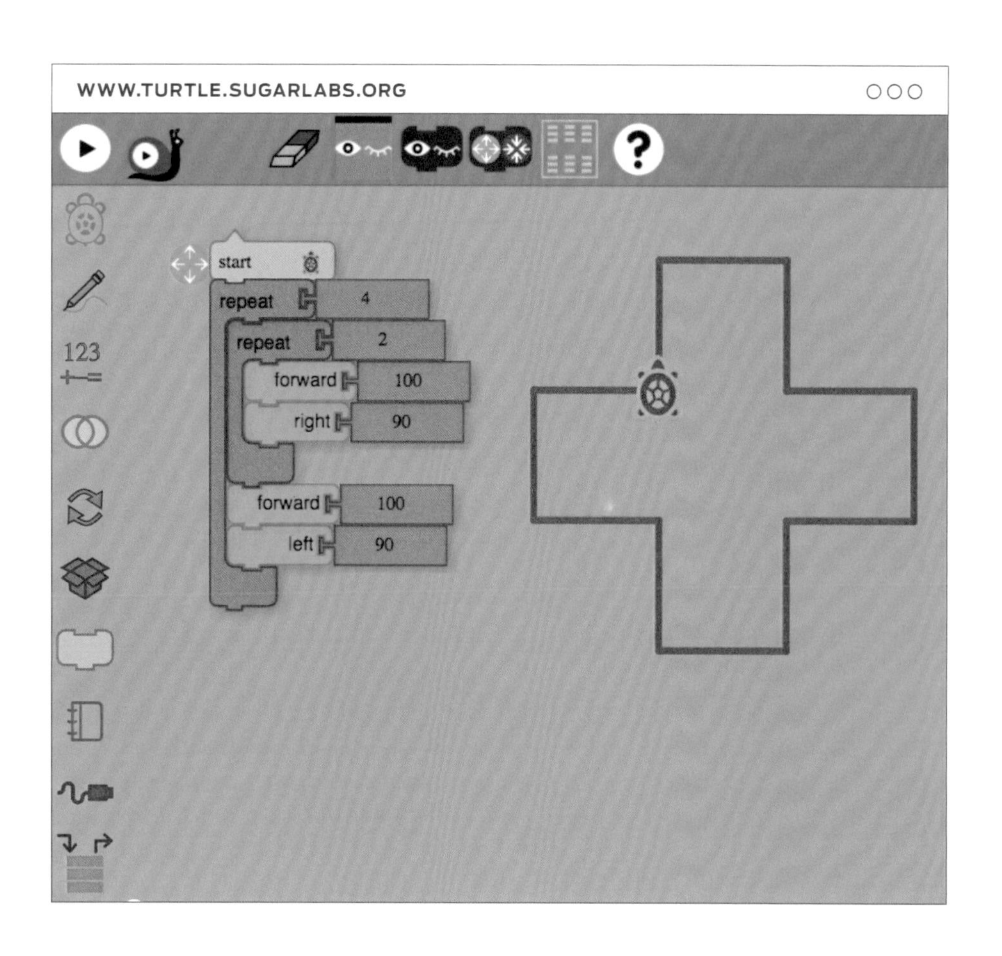

D. Make your final shape using the commands you practiced. **Check your shape** to make sure it meets your original goal. Do the lines connect to create a single object? Does it have the correct perimeter and angles?

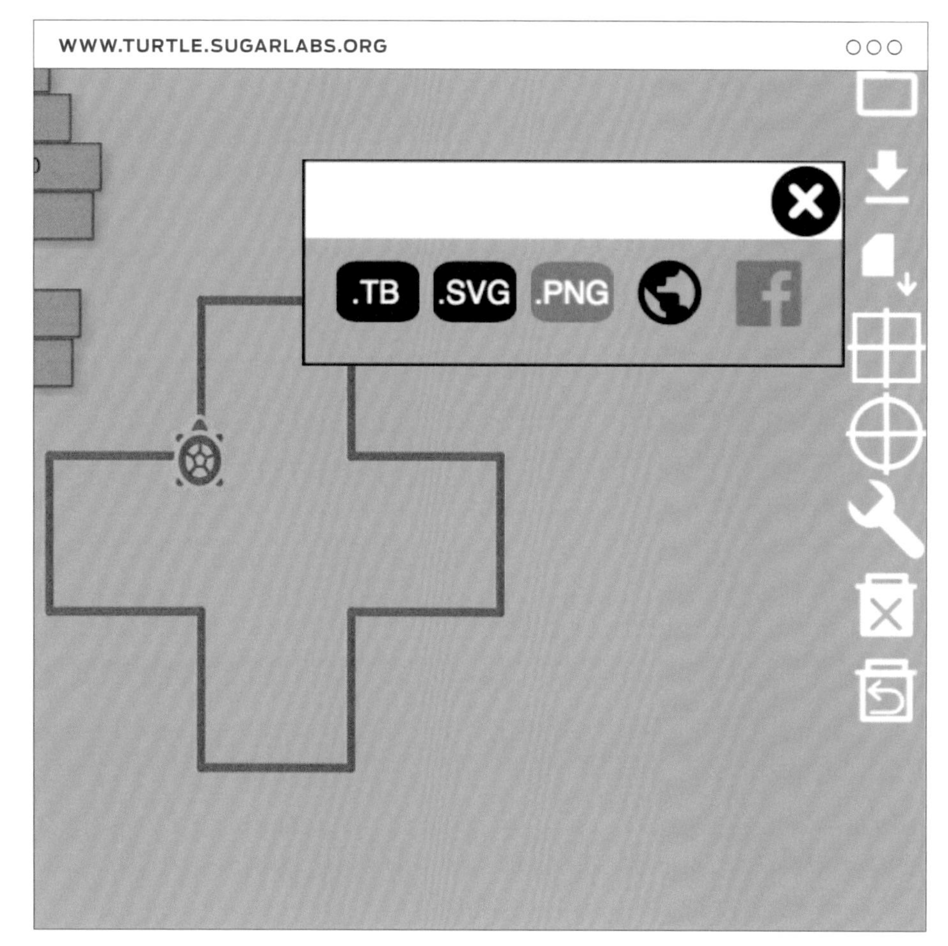

E. Use the download button on the right side of the screen to **save project**. Select .SVG and name your file. Now you're ready to import this file into Tinkercad!

STEP 03:
IMPORT YOUR SHAPE INTO TINKERCAD

A. Open Tinkercad and start a new design. Click the **import** button in the top right and select **choose a file** to import your .SVG file.

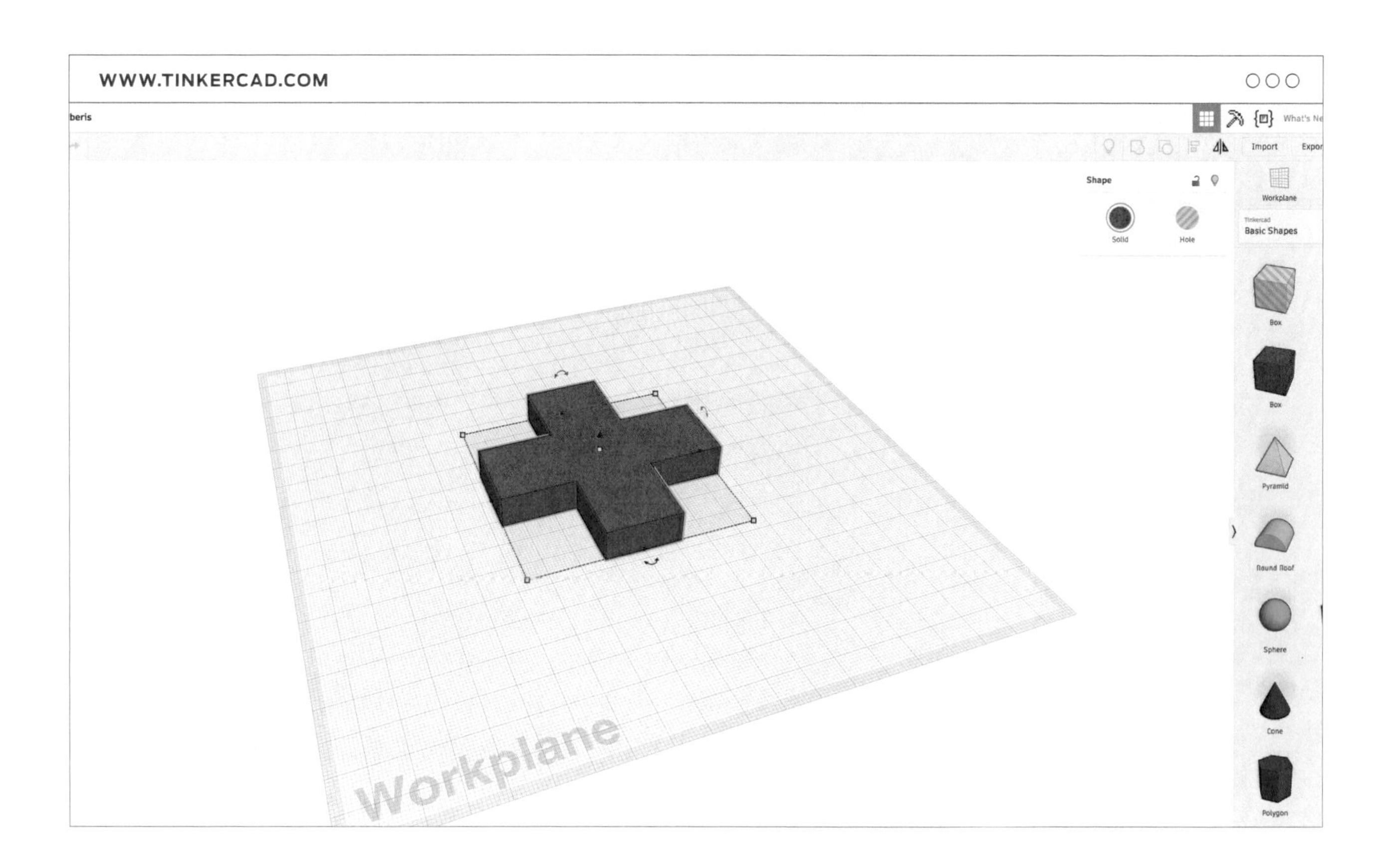

B. Alter your shape or scale however you want. Make sure you check the size with your teacher before exporting.

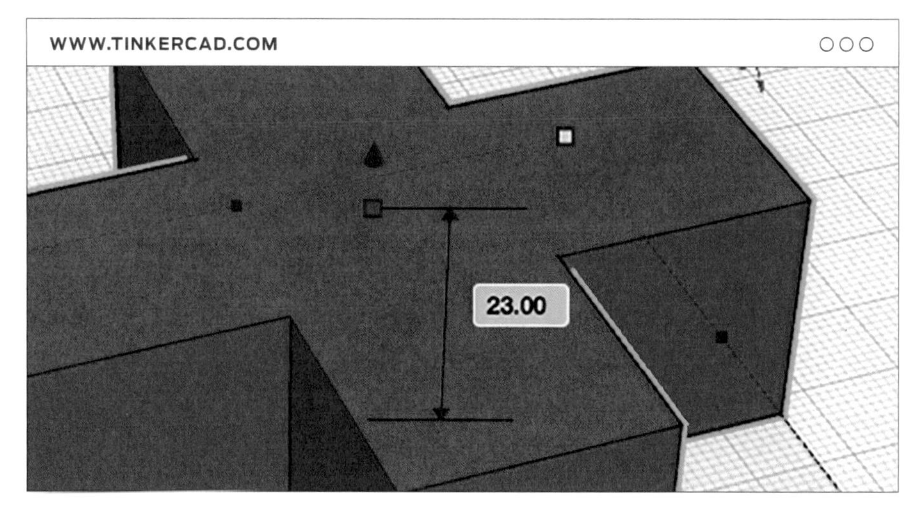

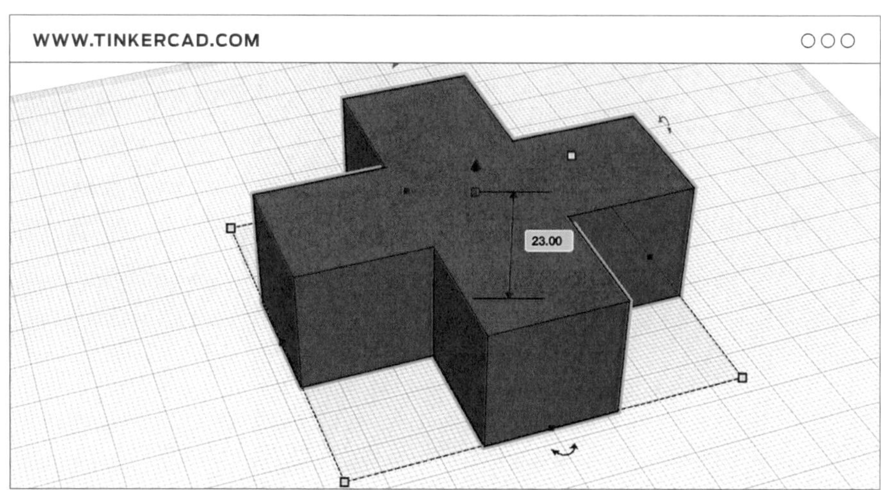

STEP 04:
PRINT YOUR SHAPE

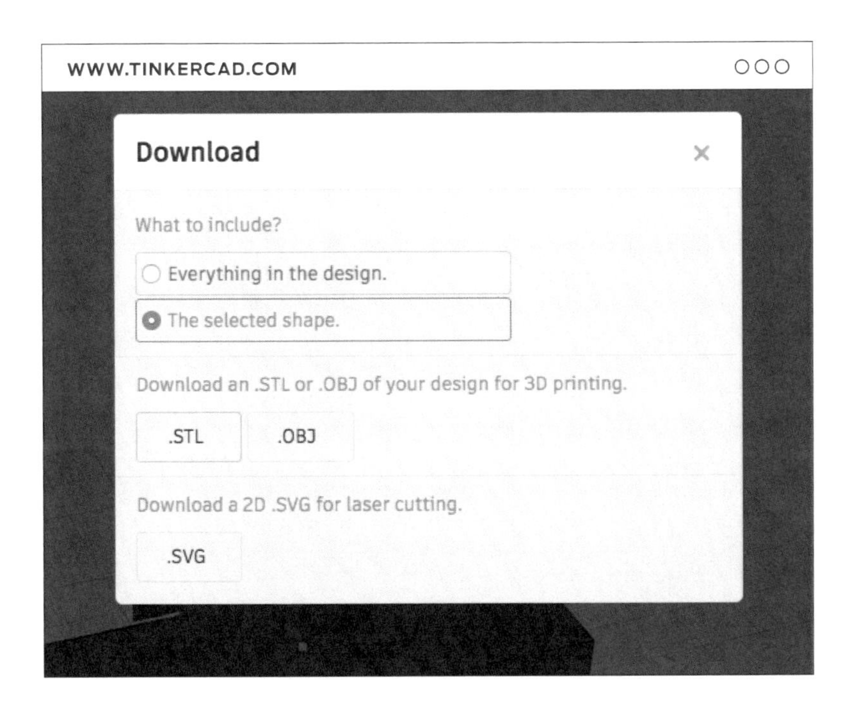

A. In Tinkercad, click export in the top right to export your file. Make sure to select the .STL option.

B. Import your file into MakerBot Print and prepare for printing. You can resize models using the scale menu, or change some of the print settings to reduce print time.

Print Settings:

Rafts	Yes
Supports	No
Resolution	0.2mm
Infill	10%

PROJECT COMPLETE:
RECODE, REPRINT, REPEAT

GOING FURTHER

A. For more advanced students, challenge them to recreate some of the complex shapes shown in the Thingiverse post.

B. For students learning trigonometry, try providing a right triangle with only 1 side length and 2 angles defined, and have students solve for the unknown angle and side lengths before creating their shape.

PROJECT FOUR

3D PRINTED MINECRAFT CASTLE TEACHES PERIMETER AND AREA

LINK: thingiverse.com/thing:1573929 for access to handouts, videos and other materials associated with this project.

PROJECT INFO

AUTHOR
Carlos Varas,
@carlosvaras

SUBJECT
Math, Technology, History

AUDIENCE
Grade Levels 3—5

DIFFICULTY
Beginner

SKILLS NEEDED
Basic Minecraft®
game experience

DURATION
5—7 Class Periods

GROUPS
15 Groups
1—2 Students / Group

MATERIALS
Graph Paper
Pencil
Ruler

SOFTWARE
Minecraft (**PC, Mac**)
Mineways (**PC, Mac**)

PRINTERS
Works with all
MakerBot® Replicator®
3D printers

PRINT TIME
2–3 hrs / per Castle

FILAMENT USED
1–2 Large Spools

"3D printing is great for teaching spacial reasoning skills, especially for perimeter and area: two major concepts in math. Given the popularity of Minecraft, this is a great lesson to bridge the gap between playing and learning. I found that my students were really interested in this lesson because they got to take their castle home!"

– **Carlos Varas**

LESSON SUMMARY

In this project, students will learn the conceptual foundation of perimeter and area by constructing their own castle in Minecraft. The learning begins with basic design principles applied in Minecraft using a medieval theme. As students plan and design their castles, they will consider concepts such as perimeter (castle moat) and area. Once printed, students can make a real connection to the difference between perimeter and area.

LEARNING OBJECTIVES

After completing this project, students will be able to:

› **Understand** the difference between perimeter and area.

› **Plan, design, and print** a 3D model using Minecraft.

› **Understand** the fundamental concept of a unit when measuring.

COMMON CORE MATH STANDARDS

3.MD.D.8. Measurement & Data Solve real world problems involving perimeters of polygons, including finding the perimeter given the side lengths, finding an unknown side length, and exhibiting rectangles with the same perimeter and different areas or with the same area and different perimeters.

4.MD.A.2. Measurement & Data Use the four operations to solve word problems involving distances, intervals of time, liquid volumes, masses of objects, and money, including problems involving simple fractions or decimals, and problems that require expressing measurements given in a larger unit in terms of a smaller unit. Represent measurement quantities using diagrams such as number line diagrams that feature a measurement scale.

4.MD.A.3. Measurement & Data Apply the area and perimeter formulas for rectangles in real world and mathematical problems.

5.MD.C.5. Measurement & Data Relate volume to the operations of multiplication and addition and solve real world and mathematical problems involving volume.

TEACHER PREPARATION

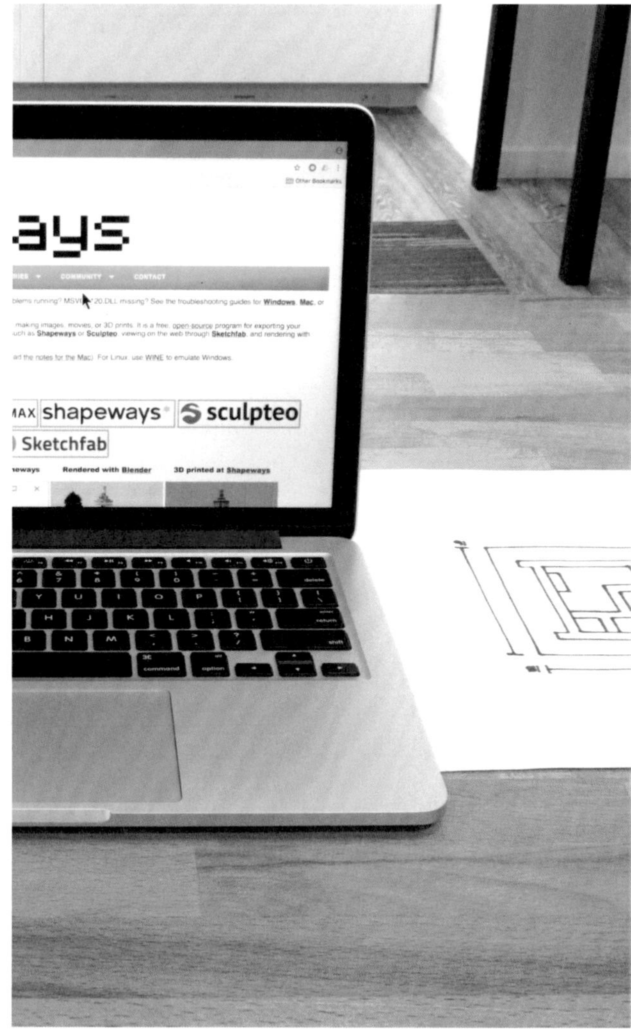

A. Download and install Minecraft on student computers. Make sure you have enough computers for each student or group.

TIP: Begin with a Minecraft Edu license if you can. You can buy a server license and a certain number of workstation licenses. If you can't do that, a regular Minecraft license will work. If you aren't using the EDU license, students will need to use the same computer each class period.

B. Download and install Mineways on each computer. This program allows you to export .STL files from Minecraft for 3D printing. It's free and pretty easy to use.

TIP: There's a tutorial video included in the Thingiverse Education post: *thingiverse.com/thing:1573929*

C. Students should be given time to explore Minecraft and how to use it. Teach your students that Minecraft uses a basic block system (1 block = 1 unit), and things in Minecraft are built using the same basic block shape. YouTube® is a great resource for getting started.

TIP: Maps are available from the Minecraft Edu Library that allow students to teleport into their own workspace so that no one can enter but them.

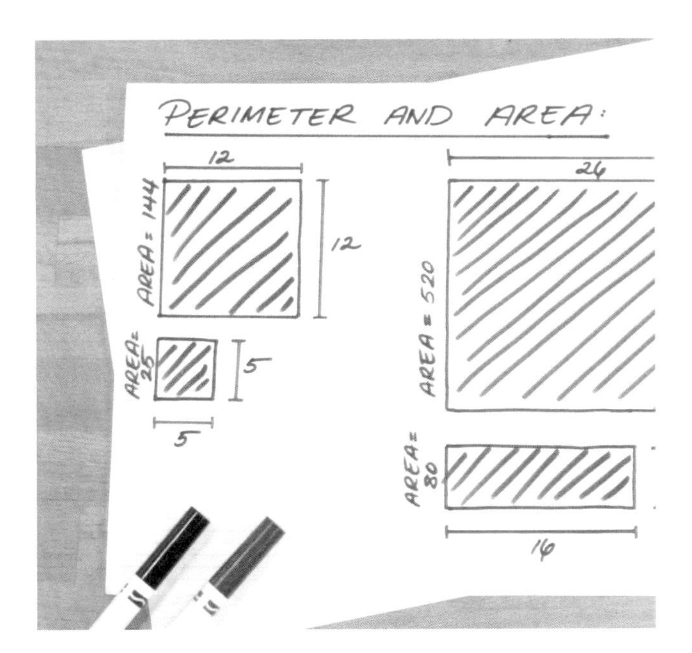

D. Discuss perimeter and area with students

› What is the difference between perimeter and area?

› Why would you want to know both the perimeter and area of an object?

› Show a few geometric shapes and point out the area and perimeter.

› Show a picture of an example Minecraft castle. Ask students to point out the castle's area and perimeter.

› Discuss the concept of a moat. How does the moat of a castle relate to the perimeter of a castle?

STUDENT ACTIVITY

STEP 01:
PLAN YOUR CASTLE

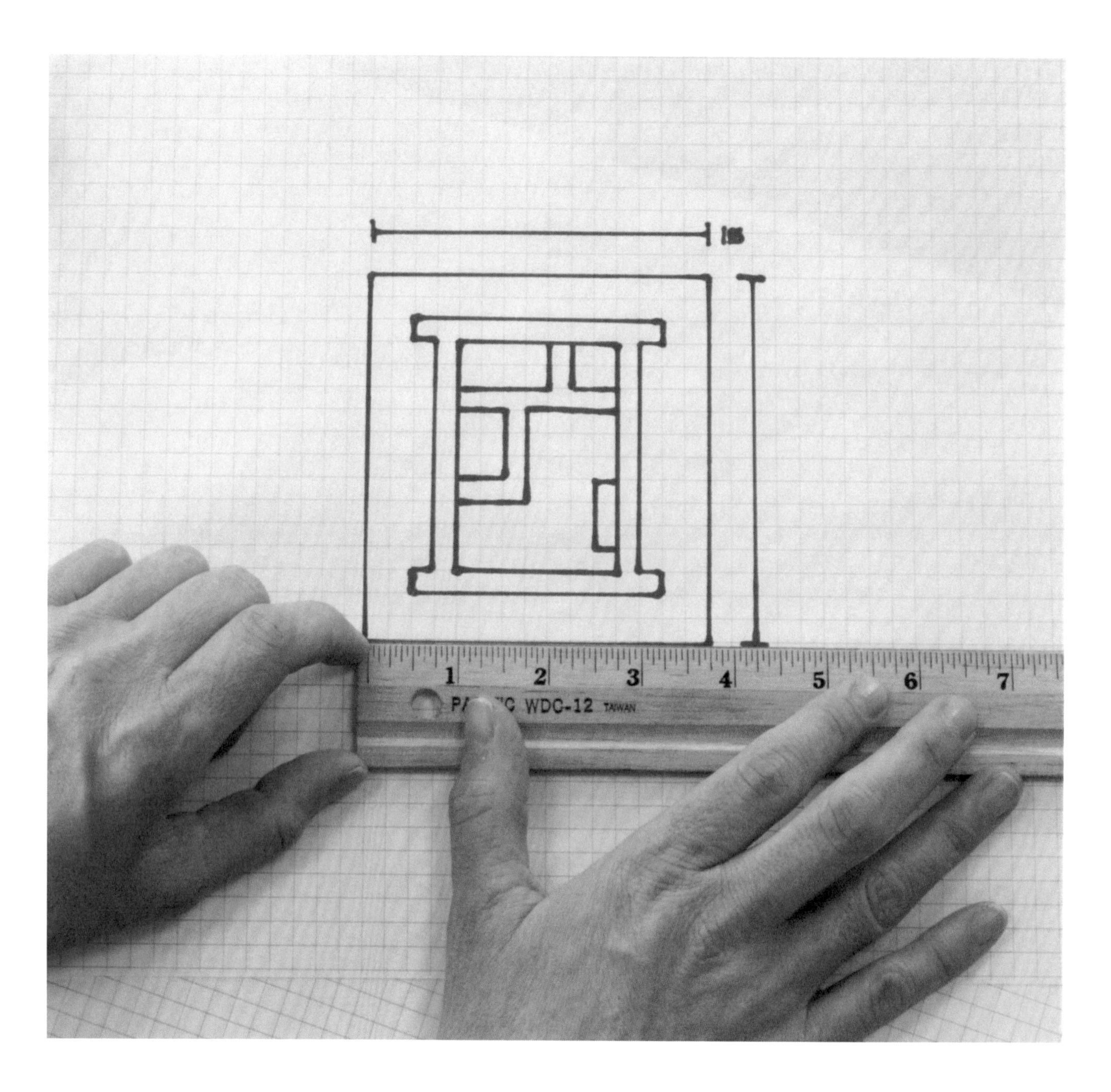

A. Research common castle layouts for inspiration on what to include in your group's castle. You can also search the Minecraft Gallery for sample Minecraft castle designs, or use the pictures from this project to help get started.

B. Plan your castle using graph paper to draw out the basic shape of your castle from an overhead view.

C. Count the unit blocks that make up the perimeter of your castle design for reference when building in Minecraft.

STEP 02:
BUILD A CASTLE IN MINECRAFT

A. Scout location: Open Minecraft Creative Mode and find a place to build your castle.

B. Start building: Be sure to build your castle with the same number of unit blocks as planned in Step 1.

C. Save your work: Save your world before leaving each class. Make sure to give your world a name so it will be easy to find later.

STEP 03:
REVIEW YOUR CASTLE DESIGN

A. Review your castle design with your teacher to make sure the walls are tall enough and the moat is deep enough to come out when you print them.

B. Remove any small details such as torches or water for easier printing.

STEP 04:
OPEN MINEWAYS AND LOCATE CASTLE

A. Open Mineways and click **File** and then click **Open World** and choose the correct world.

B. Find your castle: Once open, you'll see a bird's eye view of your world. Navigate around until you find the castle and zoom in.

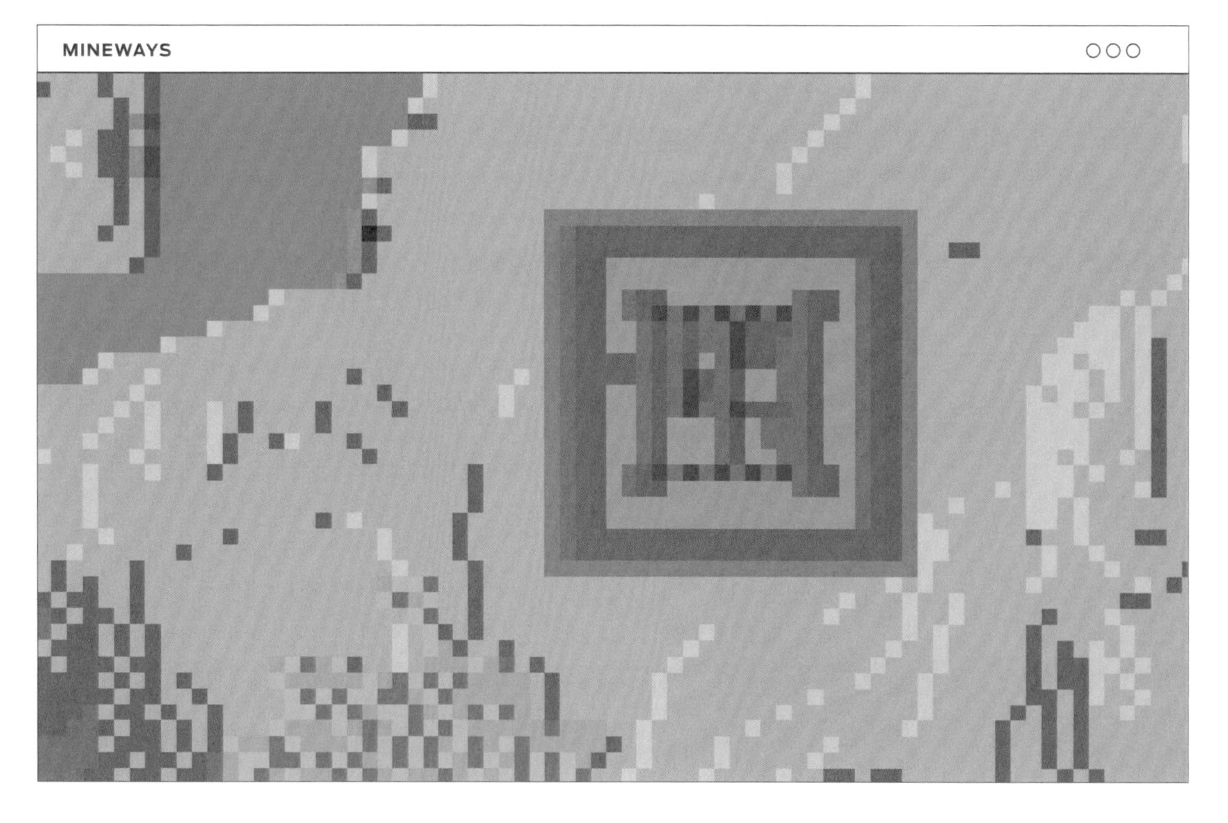

C. Select your castle: Right click and drag to select the castle. Include 1 block around it to ensure a good printing base. Your selection will be highlighted in pink.

STEP 05:
EXPORT YOUR CASTLE FROM MINEWAYS

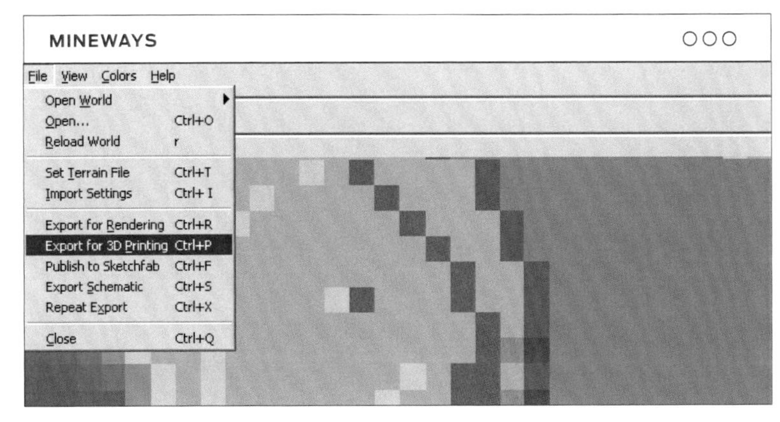

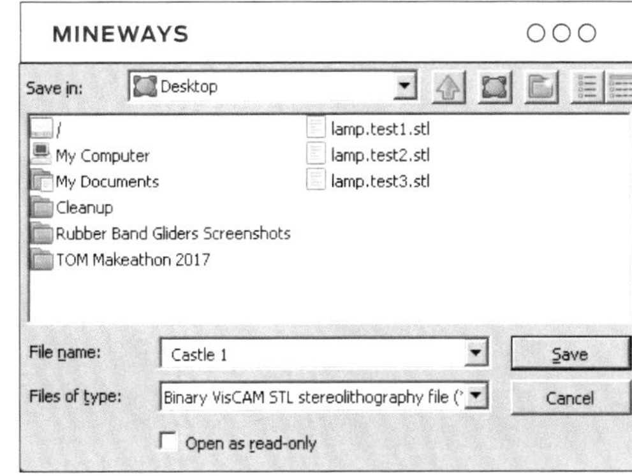

A. Export .STL: With your castle selected, click **File** and then click **export for 3D printing** and choose where to save your file. Make sure to select binary STL as the file type.

MINEWAYS

Model Export Dialog

World coordinates selection area:

Box min: X= 74 Height Y= 62 Z= 80

Box max: X= 97 Height Y= 255 Z= 102

Create a ZIP file containing all export model files / Create files themselves

Export no materials / Export solid material colors / Export richer color textures / Export full color texture patterns

OBJ file export options: Export separate types / Export individual blocks / Separate materials/blocks / Split materials into subtypes / G3D full material

Make Z the up direction instead of Y / Center model around the origin / Create block faces at the borders

Create composite overlay faces / Use biome in center of export area / Tree leaves solid (less polygons)

Rotate model clockwise: 0 90 180 270 degrees

3D printing related options:

Make the model 5 cm in height (2.54 cm per inch)

Minimize size based on wall thickness for material type

Make each block 10 mm high Physical material: Custom Printer

Aim for a cost of 25.00 Model's units: Millimeters

Fill air bubbles: Seal off entrances Fill in isolated tunnels in model's base

Connect parts sharing an edge: Connect corner tips Weld all shared edges

Delete floating objects: trees and parts smaller than 16 blocks

Hollow out bottom of model, making the walls 1.5 mm thick Superhollow

Melt snow blocks (useful with fill, seal, and hollow to make structures strong)

Export lesser, detailed blocks: Fatten lesser blocks (safer 3D printing)

Debug: show floating parts in different colors

Debug: show weld blocks in bright colors

OK Cancel

B. Check your settings: After you click **save** you will see a model export dialog open with options for formatting. You shouldn't need to change any of the default settings, just click **OK**.

STEP 06:
PRINT AND PAINT YOUR CASTLE

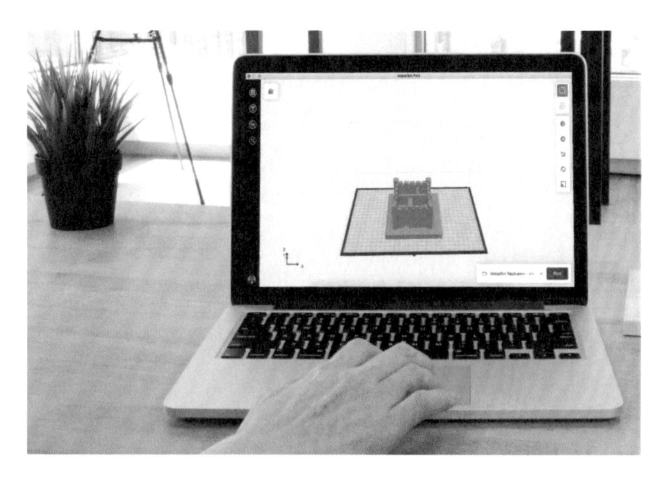

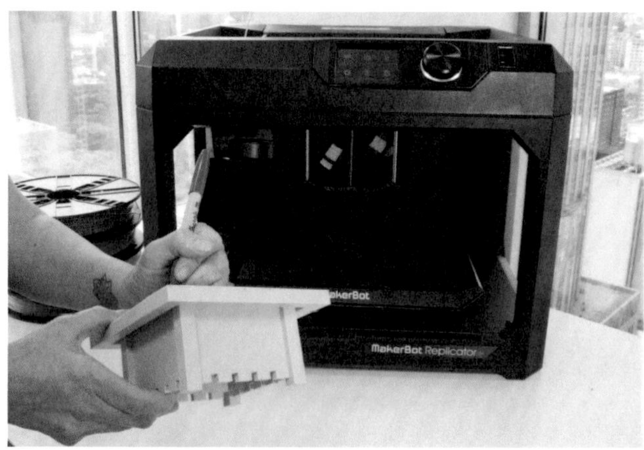

A. Import the .STL files into MakerBot Print and scale them up or down if necessary.

B. Print the designs and write your name on the bottom.

Print Settings:

Rafts	Yes
Supports	Varies
Resolution	0.2mm
Infill	10%

C. Paint and decorate your castle to bring it to life. Include some elements from your research.

PROJECT COMPLETE:
RULE YOUR KINGDOM

GOING FURTHER

A. For more advanced students, you could calculate the volume of castle and exact area if their castle isn't a perfect square or rectangle.

B. Have students create rooms inside the castle, then explain bird's eye view and basic blueprint design principles. They can also calculate the area and volume of each room individually.

PROJECT FIVE

THE SPEEDY ARCHITECT

LINK: *thingiverse.com/thing:1682347* for access to handouts, videos and other materials associated with this project.

PROJECT INFO

AUTHOR
MakerBot Learning,
@makerbotlearning

SUBJECT
Engineering,
Technology

AUDIENCE
Grade Levels 3–8

DIFFICULTY
Beginner

SKILLS NEEDED
None

DURATION
1 Class Period, 3–4 with extension activity

GROUPS
8 Groups
3–4 Students / Group

MATERIALS
Drinking Straws (50 per group)

Scissors

Ruler

3D Printed Connector Pieces (15–20 of each type per group)

Measuring tape

Two 5 lb weights

SOFTWARE
Tinkercad (**web app**)

PRINTERS
Works with all MakerBot® Replicator® 3D printers

PRINT TIME
Prep: 20 hrs / Group
Lesson: 2-3 hrs / Group

FILAMENT USED
2 Large Spools
½ Large Spool for extension activity

"This is an easy project to get started with. After printing the pieces, you can run it in a single class period. If you have a bit more time, the extension activity helps emphasize why 3D printing is such a powerful tool. Students first experience the limitations of the materials they're given, then get the chance to design and print their own custom solutions to overcome those limitations."

– **MakerBot Learning**

LESSON SUMMARY

In this quick project, students take on the role of architects to build a structure that satisfies the teacher's design requirements using a limited amount of connector pieces and straws.

Without modification, this lesson will engage students to work together in teams to overcome two challenges; building both the strongest and tallest buildings. Students will learn what shapes create the most stable buildings, how to build within design and resource constraints, and some basic 3D design skills in Tinkercad software.

This project is applicable for grades 3 - 8, and many sections of the lesson can be expanded to include more advanced engineering or design principles.

LEARNING OBJECTIVES

After completing this project, students will be able to:

› **Design** within specified requirements and restraints

› **Understand** the importance of teamwork

› **Use** Tinkercad software to modify an existing model

NGSS STANDARDS

MS-ETS1-1. Define the criteria and constraints of a design problem with sufficient precision to ensure a successful solution, taking into account relevant scientific principles and potential impacts on people and the natural environment that may limit possible solutions.

MS-ETS1-2. Evaluate competing design solutions using a systematic process to determine how well they meet the criteria and constraints of the problem.

MS-ETS1-3. Analyze data from tests to determine similarities and differences among several design solutions to identify the best characteristics of each that can be combined into a new solution to better meet the criteria for success.

TEACHER PREPARATION

A. Print the connector pieces from the Thingiverse Education project: *thingiverse.com/thing:1682347.* Print 20 - 30 of each connector per group. It will take some time to print them all, but you only need to do this step once because the pieces are all reusable for future projects. If possible, print extras of the corner connector piece, it's usually the most popular.

TIP: Print each connector piece in it's own color for easier sorting.

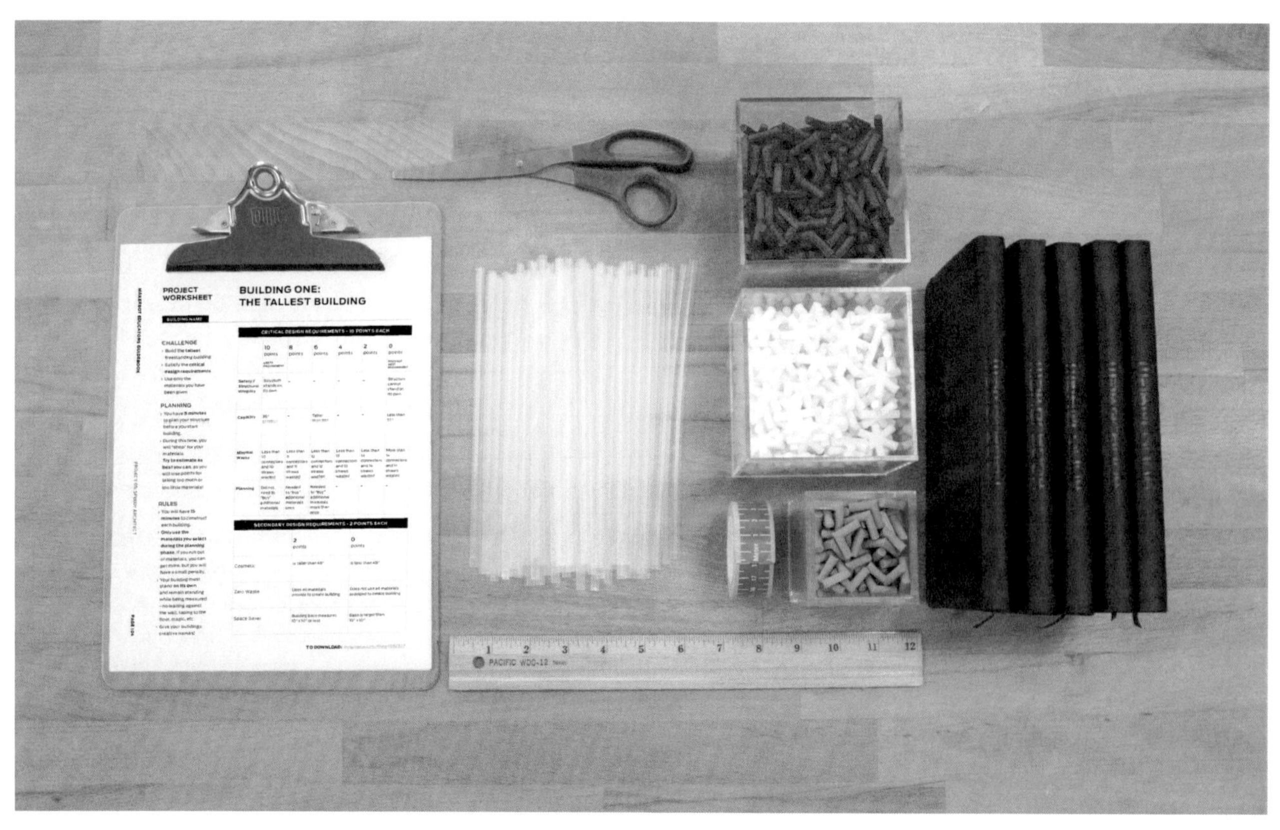

B. Lay out all building materials and student handouts. Any standard sized drinking straws with a 0.22" diameter will work. Check the link in the project's materials section on Thingiverse Education for the straws used here.

STUDENT ACTIVITY

STEP 01:

DESIGN REQUIREMENTS - THE TALLEST BUILDING

A. Get challenge handout: Get your group's Design Requirements Sheet from your teacher.

B. Review challenge: Using the materials provided, you will need to build the TALLEST possible structure that can stand on its own and remain standing for long enough to be measured. Review the critical design requirements and the secondary design requirements on your handout.

STEP 02:

PLAN AND SHOP FOR THE TALLEST BUILDING

A. Plan: Now that you've reviewed the design requirements, plan your design.

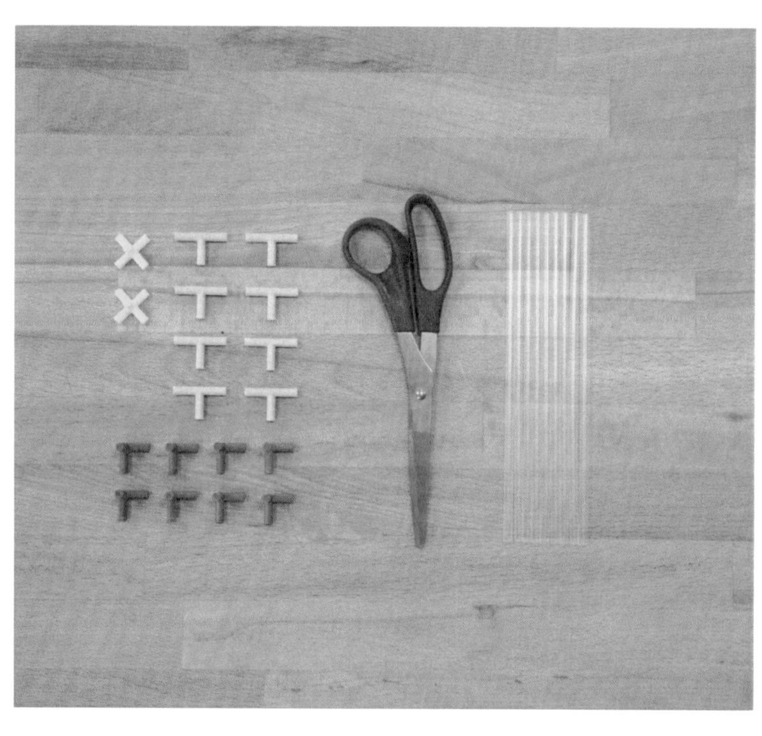

B. Shop for materials: Take a short amount of time (5 minutes or so) to shop for materials. Try to take a diverse amount of connectors, but only take what you need! This emphasizes the importance of planning and will challenge you to find creative solutions. Groups with leftover materials will receive less points.

TIP: See grading sheet about excess materials and "re-buying" materials.

STEP 03:

BUILD THE TALLEST BUILDING

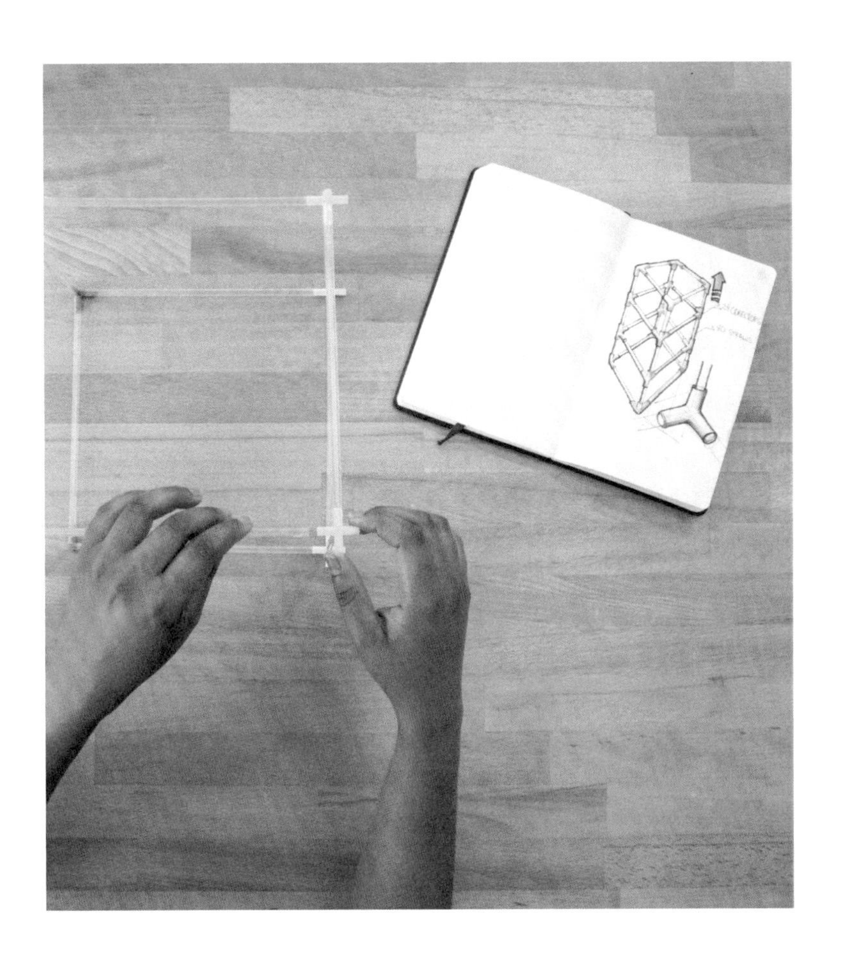

A. Build! Groups have 15 minutes to create their buildings. Reference your handout for the guidelines:

- All towers will be measured at the end of the time limit.
- You can rebuild your tower as many times as you want.
- Your tower must stand on its own and remain standing while being measured - no leaning against the wall, taping to the floor, magic, etc.
- Give your tower a creative name!

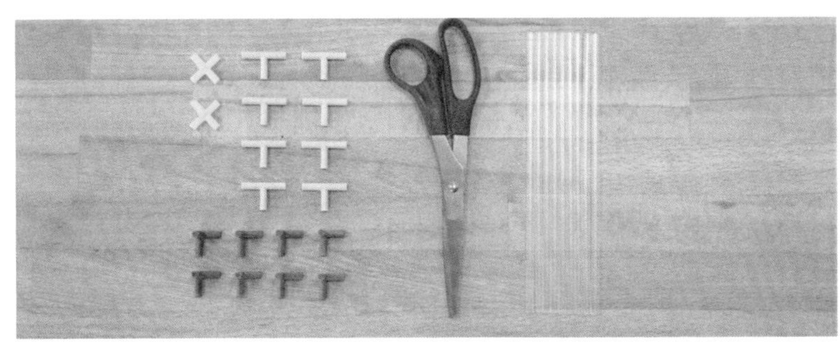

B. Measure buildings: After 15 minutes, your teacher will measure the height of your building, count your leftover materials, and record the results.

C. Disassemble: Take apart your buildings and place the materials back in the class pile to get ready for the second challenge.

STEP 04:
DESIGN REQUIREMENTS - THE STRONGEST BUILDING

A. Review challenge: Using the materials provided, build the STRONGEST possible structure that can withstand 5 lbs of weight for at least 5 seconds.

B. Design requirements: Review the critical design requirements and the secondary design requirements on your handout.

STEP 05:

PLAN AND SHOP FOR THE STRONGEST BUILDING

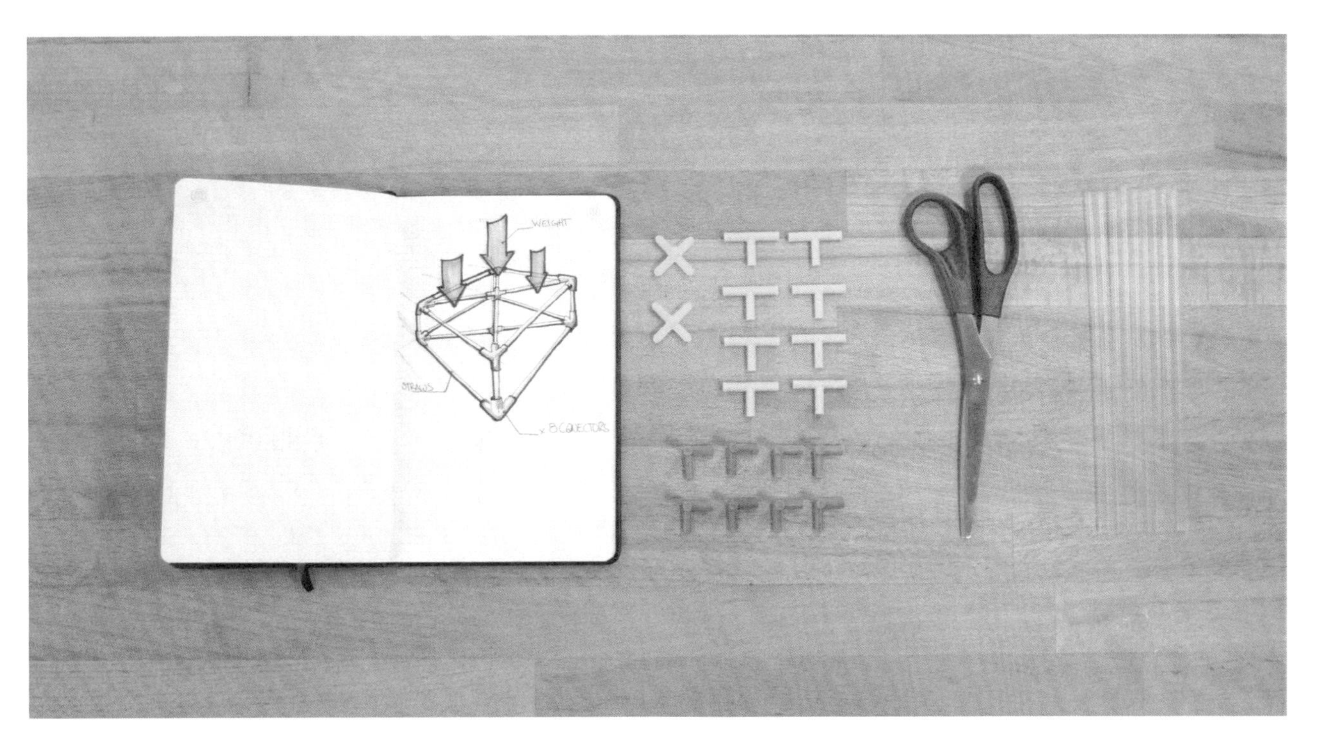

A. Plan. Now that you've reviewed the design requirements, plan your design.

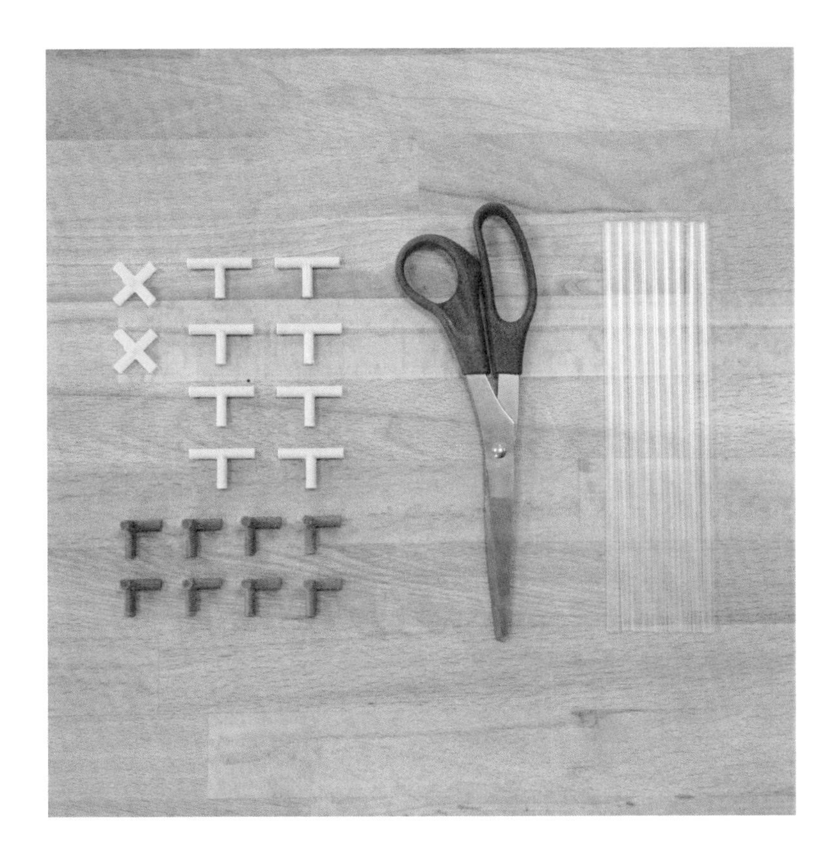

B. Shop for materials: Take a short amount of time (5 minutes or so) to shop for new materials. Try to take a diverse amount of connectors, but only take what you need! This emphasizes the importance of planning and will challenge you to find creative solutions. Groups with leftover materials will receive less points.

STEP 06:

BUILD THE STRONGEST BUILDING

A. Build! Groups have 15 minutes to create their buildings. Reference your handout for the guidelines.

B. Measure buildings: After 15 minutes, your teacher will bring a 5 lb weight to measure your building, count your leftover materials, and record the results.

C. Disassemble: Take apart your buildings and place the materials back in the class pile.

EXTENSION ACTIVITY

STEP 01:
IMPORT CONNECTOR

To add more design and engineering challenges to the project, incorporate 3D design using Tinkercad software. Design your own custom 3D printed connector using what you learned after building from a set of existing connectors. Can you improve on one of the connector designs? How would your building designs improve using a brand new 3D printed connector type?

A. Download: Start by downloading one of the connector .STL files found on the Thingiverse Education™ project page (*thingiverse.com/thing:1682347*). We used the two arm connector.

B. Navigate to Tinkercad.com and click **create new design**.

C. Import: Once you've opened a new design, navigate to the upper right hand corner of the window and select **import**. Choose your downloaded file and click **import**.

STEP 02:
RESIZE AND DUPLICATE

Because this connector was designed to hold two straws across its length, you'll need to edit the length to begin creating a new connector.

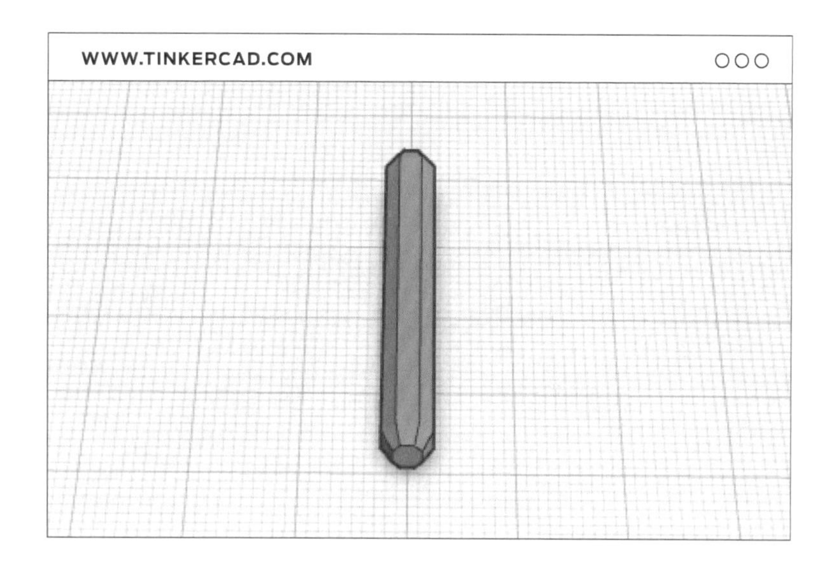

A. Insert ruler: Drag the ruler tool from the upper right hand corner of your screen onto your workplane, then select your object to see its dimensions.

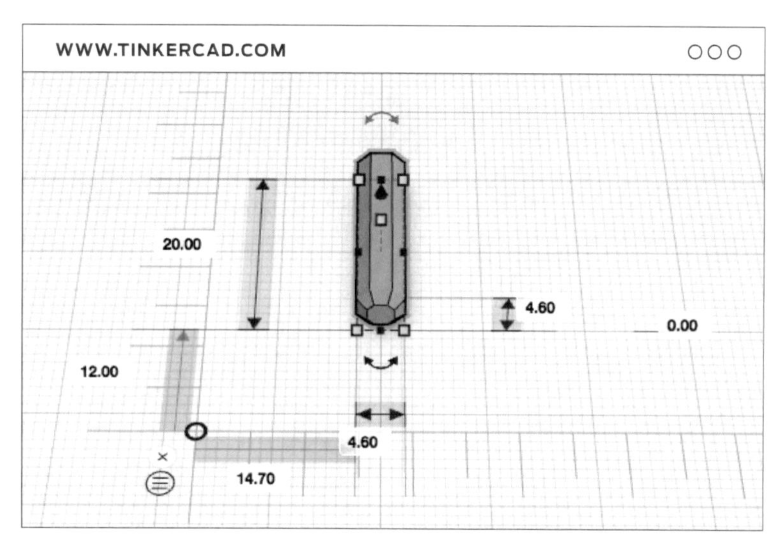

B. Resize: The connector measures 4.6 x 4.6 x 40 mm. Edit the length so that the dimensions are 4.6 x 4.6 x 20 mm. Now that the connector leg is the proper size, you can duplicate it to create new custom connectors with several legs.

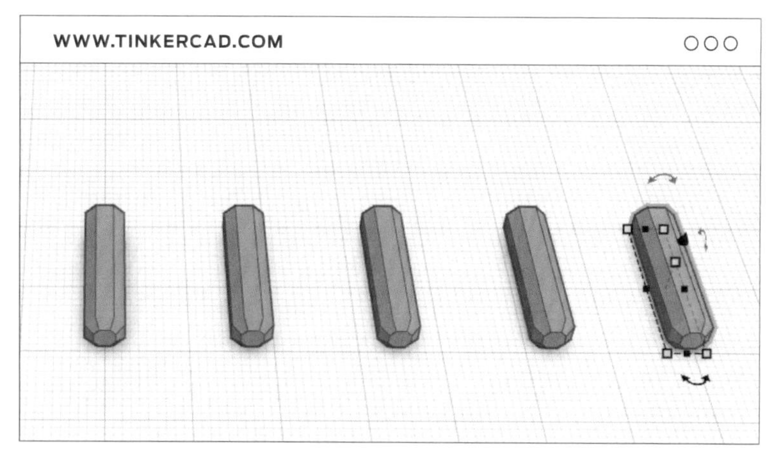

C. Copy and paste: To copy a part, press **ctrl-C** then **ctrl-V** (**cmd-C**, **cmd-V** on Mac). Create several copies depending on the number of legs you want your connector to have.

STEP 03:
BUILD A NEW CONNECTOR

Now that you've duplicated your connector legs, it's time to design an all new connector by moving and rotating the legs created into place.

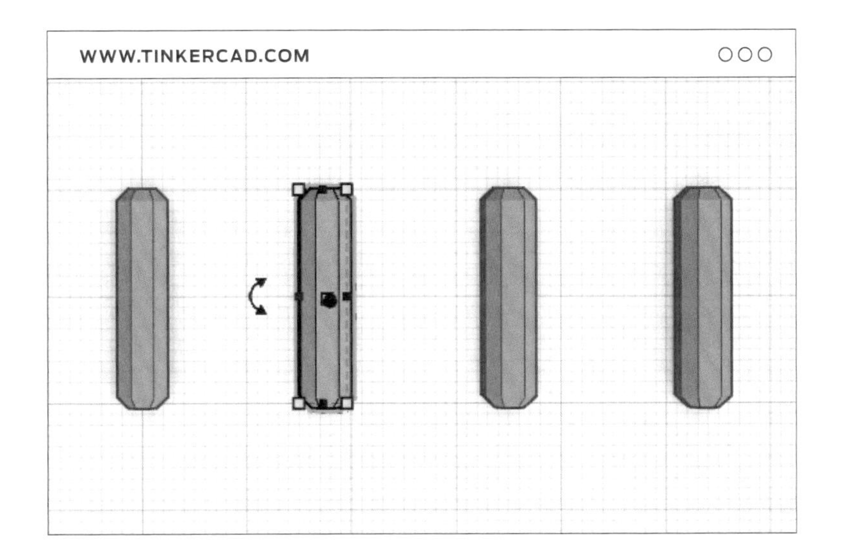

A. Move and rotate: Move each leg into place to build a custom connector. You can create more duplicates if needed.

TIP: The rotate tool appears around your part when selected. If you hold **shift** while rotating, your object will rotate in 45° increments, making repositioning connector legs a breeze.

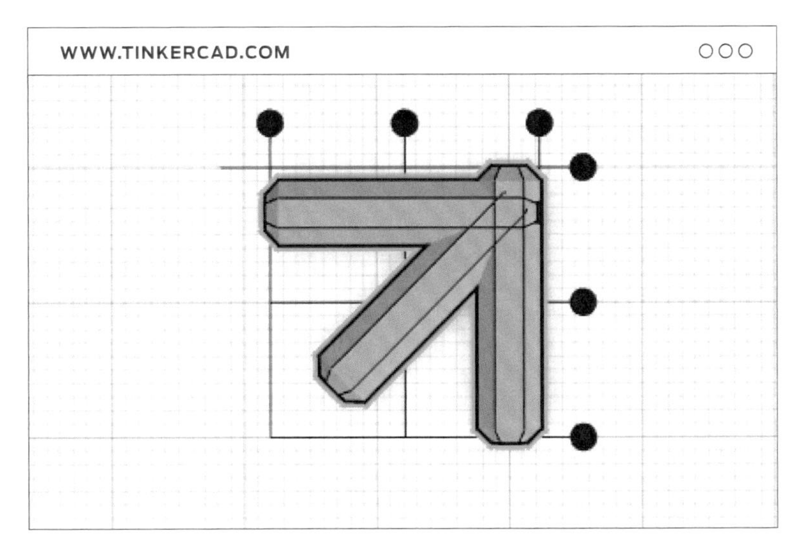

B. Align: Once you have all of your legs in place, it's best to use the align tool to make sure they're in the right place. Select 2 or more parts, then click **align** in the upper right hand corner. Click on the circles that appear to align parts to the center, left, or right on each axis.

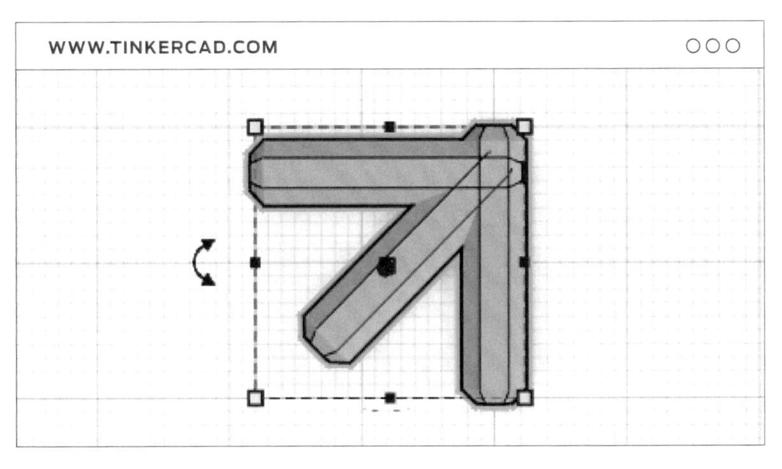

C. Group: Select all legs of your connector by holding **shift** and clicking on each leg. Then click the **group** in the upper right hand corner. Once your model has been grouped, you're ready to print!

STEP 04:

EXPORT AND PRINT

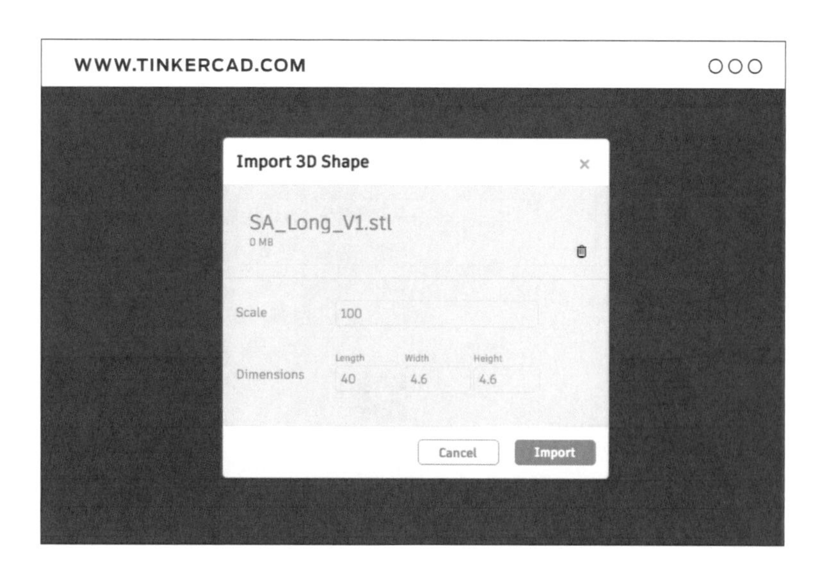

A. Export: Click **export** in the upper right hand corner of the screen and choose to download your new connector as an .STL file.

B. Import: Import your new connector in MakerBot Print and re-orient the part if needed. If you want to print multiple, right click on your model, select **duplicate**, and enter the number of copies you want.

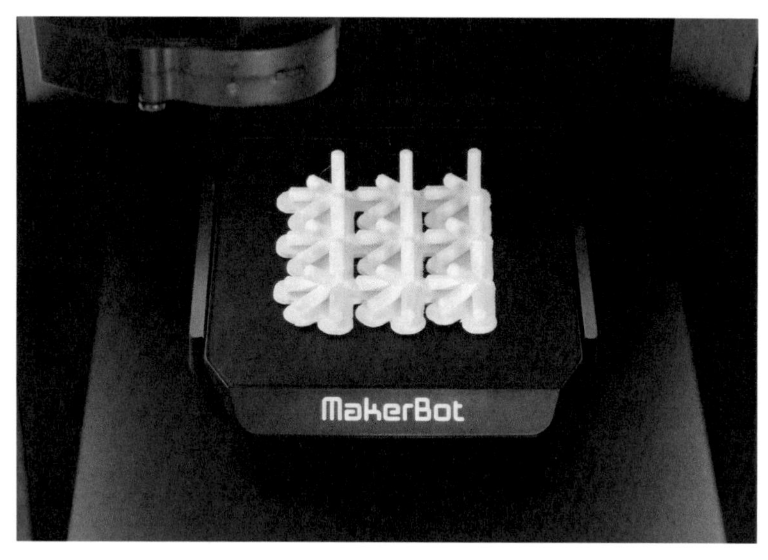

C. Print: Select your print settings and begin printing!

Print Settings:

Rafts	Yes
Supports	No
Resolution	0.2mm
Infill	10%

PROJECT COMPLETE:
BUILD FOR TOMORROW

GOING FURTHER

A. Discuss the following reflection questions:

- What was your initial design plan?
- How did this initial design change as you built your buildings?
- Did you ever start over?
- What was the final height of your Tall Building? How much weight could your Strong Building hold?
- What was the hardest part of building your structures?
- Was there anything else you learned while building?

PROJECT WORKSHEET

BUILDING ONE: THE TALLEST BUILDING

BUILDING NAME ____________________

CHALLENGE

- Build the **tallest** freestanding building
- Satisfy the **critical design requirements**
- Use only the materials you have been given

PLANNING

- You have **5 minutes** to plan your structure before you start building.
- During this time, you will "shop" for your materials. **Try to estimate as best you can**, as you will lose points for taking too much or too little materials!

RULES

- You will have **15 minutes** to construct each building.
- **Only use the materials you select during the planning phase**. If you run out of materials, you can get more, but you will have a small penalty.
- Your building must stand **on its own** and remain standing while being measured - no leaning against the wall, taping to the floor, magic, etc.
- **Give your buildings creative names!**

CRITICAL DESIGN REQUIREMENTS - 10 POINTS EACH

	10 points MEETS REQUIREMENT	**8** points	**6** points	**4** points	**2** points	**0** points DOES NOT MEET REQUIREMENT
Safety / Structural Integrity	Structure stands on its own.	-	-	-	-	Structure cannot stand on its own.
Capacity	36" or taller	-	Taller than 30"	-	-	Less than 30"
Minimal Waste	Less than 10 connectors and 10 straws wasted	Less than 11 connectors and 11 straws wasted	Less than 12 connectors and 12 straws wasted	Less than 13 connectors and 13 straws wasted	Less than 14 connectors and 14 straws wasted	More than 14 connectors and 14 straws wasted
Planning	Did not need to "buy" additional materials	Needed to "buy" additional materials once	Needed to "buy" additional materials more than once	-	-	-

SECONDARY DESIGN REQUIREMENTS - 2 BONUS POINTS EACH

	2 bonus	**0** bonus
Cosmetic	Is taller than 48"	Is less than 48"
Zero Waste	Uses all materials provide to create building	Does not use all materials provided to create building
Space Saver	Building base measures 10" x 10" or less	Base is larger than 10" x 10"

TO DOWNLOAD: thingiverse.com/thing:1682347

PROJECT WORKSHEET

BUILDING TWO: THE STRONGEST BUILDING

BUILDING NAME

CHALLENGE

- Build the **strongest** freestanding building
- Satisfy the **critical design requirements**
- Use only the materials you have been given

PLANNING

- You have **5 minutes** to plan your structure before you start building.
- During this time, you will "shop" for your materials. **Try to estimate as best you can**, as you will lose points for taking too much or too little materials!

RULES

- You will have **15 minutes** to construct each building.
- **Only use the materials you select during the planning phase**. If you run out of materials, you can get more, but you will have a small penalty.
- Your building must stand **on its own** and remain standing while being measured - no leaning against the wall, taping to the floor, magic, etc.
- **Give your buildings creative names!**

CRITICAL DESIGN REQUIREMENTS - 10 POINTS EACH

	10 points MEETS REQUIREMENT	8 points	6 points	4 points	2 points	0 points DOES NOT MEET REQUIREMENT
Safety / Structural Integrity	Structure stands on its own.	-	-	-	-	Structure cannot stand on its own.
Strength	Can hold a 5lb weight for 5 secs without breaking	Can hold a 5lb weight for 4 secs without breaking	Can hold a 5lb weight for 3 secs without breaking	Can hold a 5lb weight for 2 secs without breaking	Can hold a 5lb weight for 1 sec without breaking	Cannot hold a 5lb weight without breaking
Minimal Waste	Less than 10 connectors and 10 straws wasted	Less than 11 connectors and 11 straws wasted	Less than 12 connectors and 12 straws wasted	Less than 13 connectors and 13 straws wasted	Less than 14 connectors and 14 straws wasted	More than 14 connectors and 14 straws wasted
Planning	Did not need to "buy" additional materials	Needed to "buy" additional materials once	Needed to "buy" additional materials more than once	-	-	-

SECONDARY DESIGN REQUIREMENTS - 2 BONUS POINTS EACH

	2 bonus	0 bonus
Super Strength	Can hold a 10lb weight without breaking (**bending okay**)	Cannot hold a 10lb weight without breaking
Zero Waste	Uses all materials provided	Does not use all materials provided
Space Saver	Building base measures 10" x 10" or less	Base is larger than 10" x 10"

TO DOWNLOAD: thingiverse.com/thing:1682347

PROJECT SIX

H_2MAKE IT GO

LINK: thingiverse.com/thing:1622569 for access to handouts, videos and other materials associated with this project.

PROJECT INFO

AUTHORS
Amy Otis, @*otisa*
Kurt Knueve, @*knuevek*

SUBJECT
Engineering, Technology

AUDIENCE
Grade Levels 4–8

DIFFICULTY
Intermediate

SKILLS NEEDED
Basic Tinkercad software experience

DURATION
7–8 Class Periods

GROUPS
4 Groups
6–7 Students / Group

MATERIALS
Sphero® robot

Tablet or Mobile Device with Sphero app

Measuring Cup

Small Cups or Containers to hold 100 mL of water

Track Material (tape, meter sticks, etc.)

Stopwatch or Timer

Ruler or Calipers

Optional: Pipe Cleaners, Rubber Bands

SOFTWARE
Tinkercad (**web app**)

PRINTERS
Works with all MakerBot® Replicator® 3D printers

PRINT TIME
Prep: 2.5 hrs / Hat
Lesson: 4 hrs / Group

FILAMENT USED
1–1 ½ Large Spool

"This lesson is a great depiction of combining multiple disciplines into one project. The students were so excited with the programming and racing that they forgot how much planning and math they were doing in the process!"

– **Amy Otis**

– **Kurt Knueve**

LESSON SUMMARY

Students will be asked to create a vehicle that attaches to the Sphero app-enabled robot using the given hat. This vehicle must be able to carry 100 mL of water from the starting line, around a track and to the finish line. Students should aim to spill as litle water as possible while speeding around the track.

LEARNING OBJECTIVES

After completing this project, students will be able to:

› **Understand** volume and how much space is required to hold a specific volume of water.

› **Design** around real-world constraints (such as size of cup and hat attachment).

› **Use** Tinkercad software to design elements necessary to help the car drive smoother, keep water in place, and prevent splashing.

› **Understand** how weight, materials, and other factors affect speed and efficiency.

NGSS STANDARDS

4-PS3-3 Energy Ask questions and predict outcomes about the changes in energy that occur when objects collide.

4-PS3-4 Energy Apply scientific ideas to design, test, and refine a device that converts energy from one form to another.

3-5-ETS1-1 Engineering and Design Define a simple design problem reflecting a need or a want that includes specified criteria for success and constraints on materials, time, or cost.

3-5-ETS1-2 Engineering and Design Generate and compare multiple possible solutions to a problem based on how well each is likely to meet the criteria and constraints of the problem.

3-5-ETS1-3 Engineering and Design Plan and carry out fair tests in which variables are controlled and failure points are considered to identify aspects of a model or prototype that can be improved.

COMMON CORE MATH STANDARDS

4.MD.A.2 Measurement and Data Solve word problems involving distances, intervals of time, liquid volumes, masses of objects, and money, including problems involving simple fractions or decimals, and problems that require expressing measurements given in a larger unit in terms of a smaller unit.

5.MD.C.3 Measurement and Data Recognize volume as an attribute of solid figures and understand concepts of volume measurement.

5.MD.C.5 Measurement and Data: Relate volume to the operations of multiplication and addition and solve real world and mathematical problems involving volume.

TEACHER PREPARATION

A. Print the sample files from Thingiverse Education™ (hat, vehicle, wheels, and axles) and assemble the vechicles to demonstrate the concept to the students.
www.thingiverse.com/thing:1622569

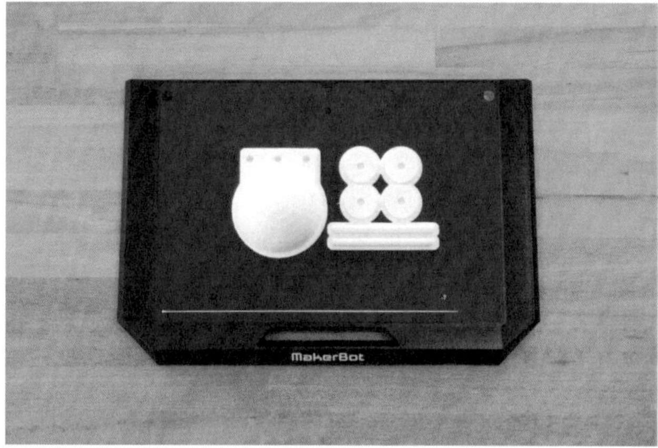

B. Print one hat file for each group. This way you'll only need to print vechicles later on.

TIP: Remind the students that these are just some examples and not necessarily the best. It's their job to come up with more creative solutions!

C. Set up your Sphero: We used the Sphero® Sprk+®, the educational, programmable version. Go to Sphero.com for help getting started.

D. Download the Sphero app and pair Sphero(s) with your tablet or mobile device(s). Students should be given a chance to become familiar with the Sphero control and its driving capabilities before starting the project.

E. Lay out the track on the floor of your classroom using tape or other materials. Make sure to identify a start and end line to make timing easier.

STUDENT ACTIVITY

STEP 01:
CREATE TINKERCAD SOFTWARE ACCOUNT

A. Create an account for your group for design purposes. If you're unfamiliar with Tinkercad software, work with the program prior to beginning this project to practice design skills.

STEP 02:
DESIGN VEHICLES

Each group should design their own vehicle using Tinkercad software then test it.
Your vehicle should attach to the Sphero hat provided and should be able to carry a container with 100 mL of water.

A. Measure the water cup and Sphero hat using a ruler or calipers. This way you can ensure your vehicle will be able to properly attach to the Sphero and carry the water cup.

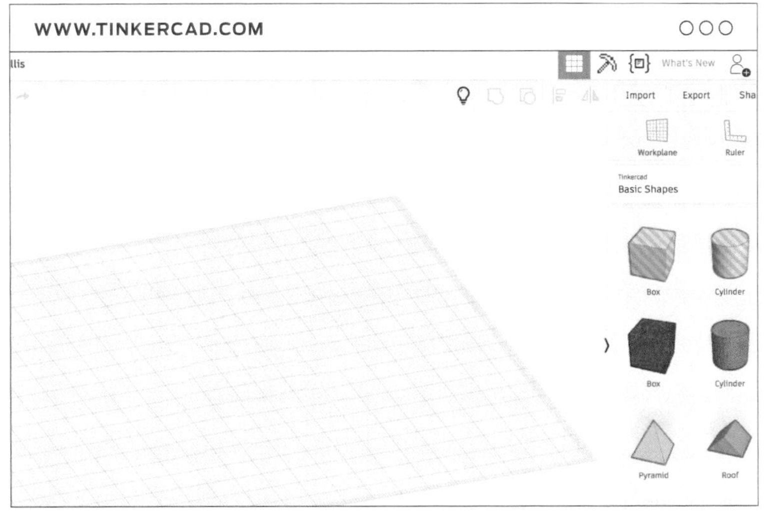

B. Start a new design in Tinkercad software.

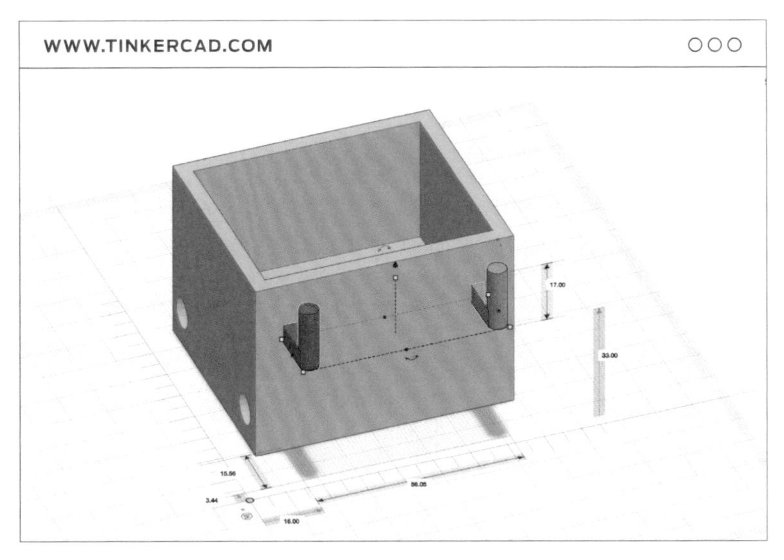

C. Use the basic shapes, holes, and group tool to create your vehicle. The **ruler** is helpful to make sure your vehicle and components are the correct size.

D. Check your work: Your vehicle should have the following components:

› A place to hold the water cup

› Properly placed prongs for attaching to the hat

› Correctly sized holes for axles

› Wheels and axles

TIP: You can either start by importing the sample files included on Thingiverse® or start from scratch.

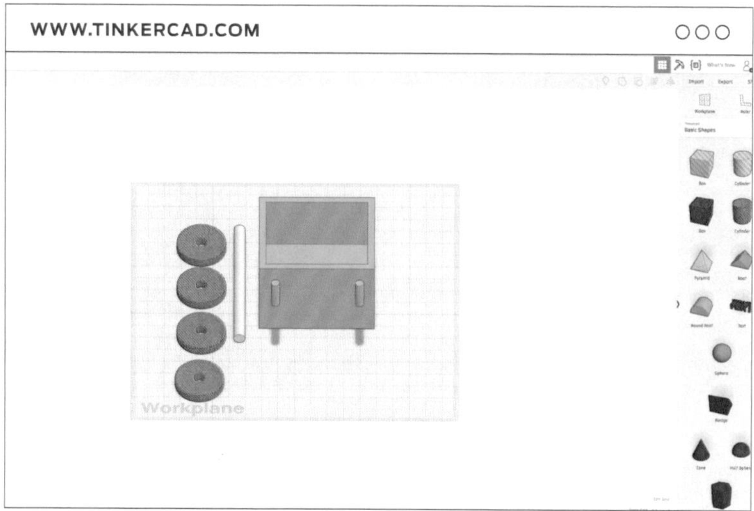

TIP: If students import the Sphero hat .STL into Tinkercad software, they can make sure their prongs line up before printing.

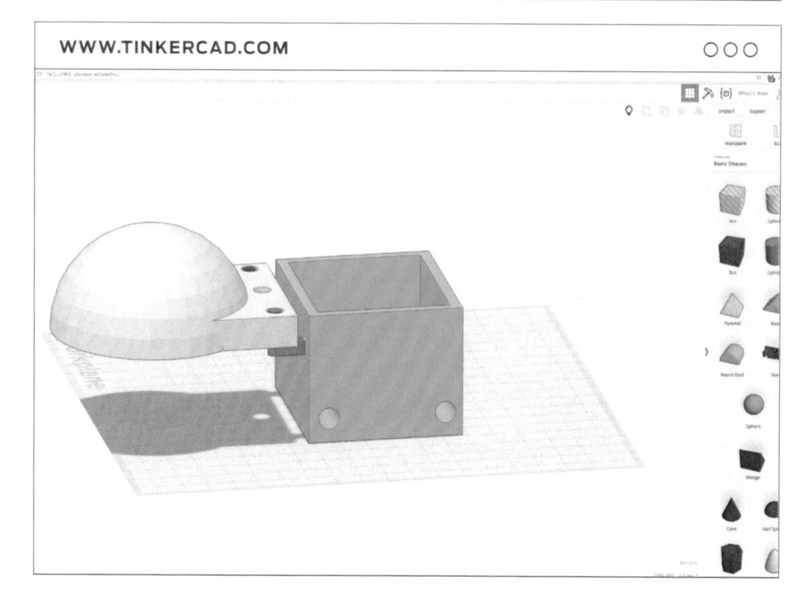

STEP 03:
PRINT, TEST, AND ITERATE!

A. Export models from Tinkercad software using the **export** button in the top right. Choose .STL format to download the files and design aspects.

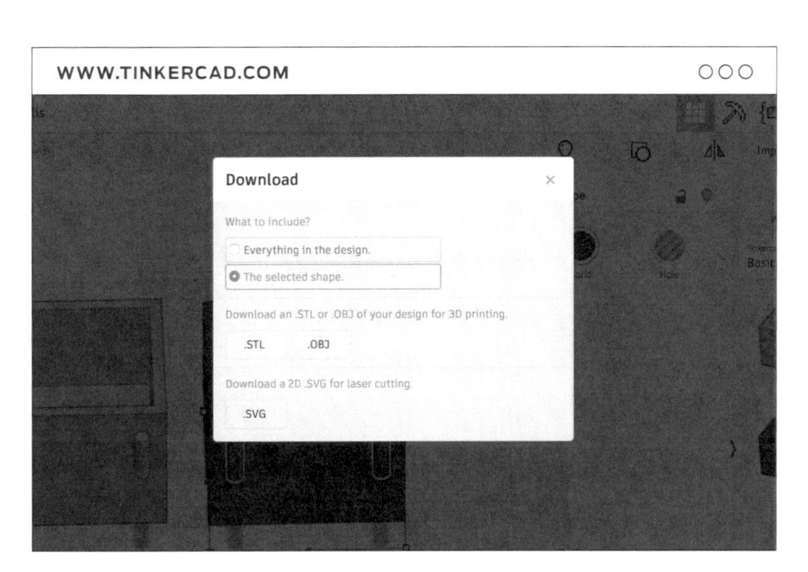

TIP: Select **download the selected shape** to export parts of your model individually.

B. Import your files into MakerBot Print and prepare for printing. The sample vehicle is able to print without support material, but student designs may need support material.

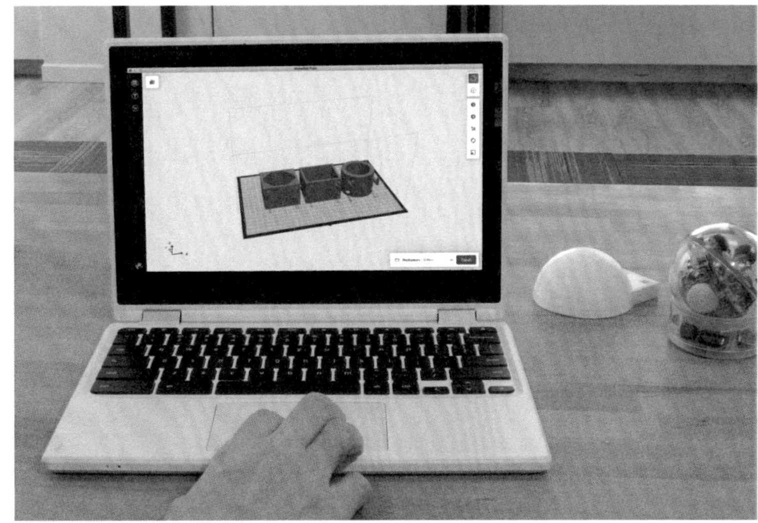

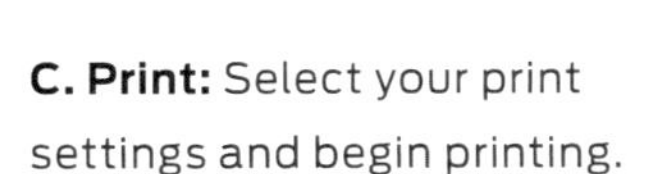

C. Print: Select your print settings and begin printing.

Print Settings:

Rafts	Yes
Supports	Varies
Resolution	0.2mm
Infill	5%

D. Assemble your vehicle: Make sure your wheels and axles attach properly and can roll.

E. Attach hat: Make sure the vehicle attaches to the hat. You can use pipe cleaners or rubber bands to help.

F. Insert cup and test: Make sure the water cup fits into your vehicle and that the Sphero can still drive with the vehicle attached.

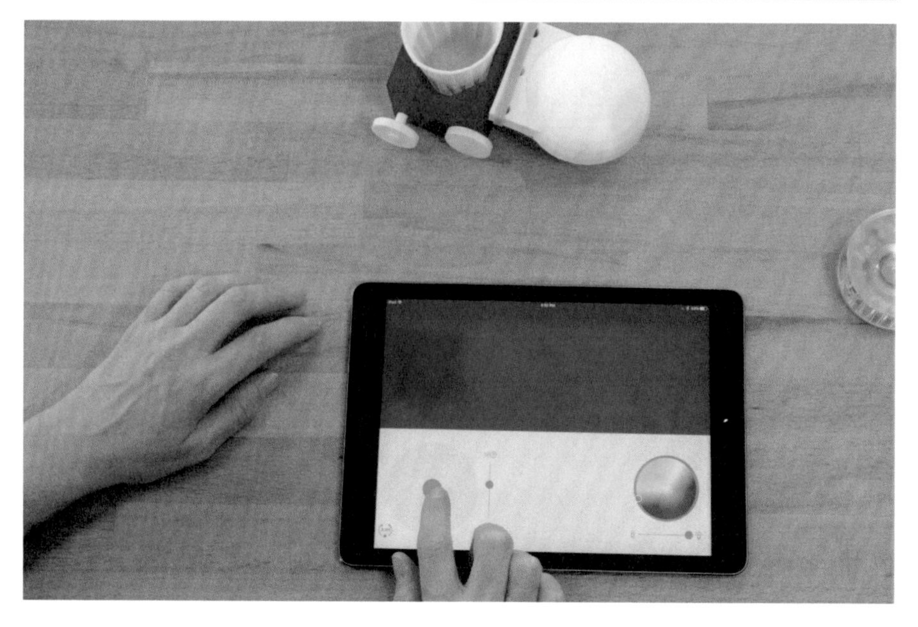

PROJECT COMPLETE:
START YOUR ENGINES!

A. Race! The goal of the race is to transport 100 mL of water from the starting line, around the track and to the finish line. You should aim to save as much water as possible while navigating the track in the least amount of time.

B. Measure the exact amount of water in your cup before starting, then measure the cup after racing. Subtract your final amount from the starting amount to figure out how much water you lost during the race, then record that value.

C. Record your time

D. Discuss: Compare class results and discuss which designs were most/least successful and why.

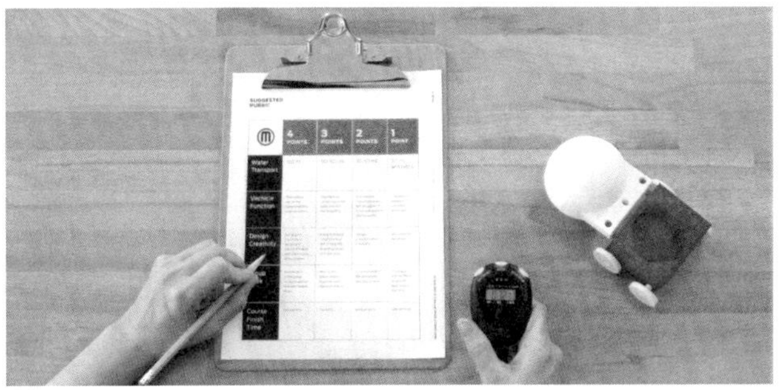

PROJECT RUBRIC

H_2MAKE IT GO

GROUP NAME ______________________

	4 points	3 points	2 points	1 points	points earned
Water Transport	100mL	90-80mL	70-60mL	50mL and below	
Function	The vehicle carries the water smoothly around corners.	The vehicle carries the water fairly well but not smoothly.	The vehicle carries the water, but struggles on turns and does not ride smoothly.	The vehicle does not carry the water well.	
Creativity	Group used creativity in design and overall it helped in success of the project.	Design showed creativity that didn't help the overall success of the project.	Design shows minimal creativity.	No creativity was used.	
Group Work	All members of the group worked together and contributed ideas.	Most of the group worked together with few distractions.	Less than half of the group was working on task.	The group did not finish or was off task most of the time.	
Course Completion Time	Fastest time	Top 50%	Bottom 50%	Did not finish	
				TOTAL POINTS	

TO DOWNLOAD: thingiverse.com/thing:1622569

GOING FURTHER

A. Try having the students use the Sphero app to code directions for the Sphero to navigate the track on its own.

PROJECT
SEVEN

WEATHER SURVIVAL CHALLENGE

LINK: *thingiverse.com/thing:1711704* for access to handouts, videos and other materials associated with this project.

PROJECT INFO

AUTHOR
Nichole Thomas,
@liberty3D

SUBJECT
Science

AUDIENCE
Grade Levels 2–6

DIFFICULTY
Intermediate

SKILLS NEEDED
Basic Tinkercad
software experience

DURATION
2-3 Class Periods

GROUPS
10 Groups
2-3 Students / Group

MATERIALS
Plastic tub

Sand or stones

Water

Sketchbook

SOFTWARE
Tinkercad (**web app**)

PRINTERS
Works with all
MakerBot® Replicator®
3D printers

PRINT TIME
Prep: None
Lesson: 5-10 hrs / Group

FILAMENT USED
1-2 Large Spools

"I love this project because it covers so many different lessons and strategies. When testing their designs, the students not only engage in the basics of STEM education, but also learn how to design (and redesign) and make the best model for the job. They get a chance to try, fail, redesign and try again – something that many students don't get the opportunity to do. It's awesome seeing all of the different solutions that the students create to solve the same problem!"

– **Nichole Thomas**

LESSON SUMMARY

This project is a great expansion for any class unit discussing weather patterns and how they affect us. Students are challenged to create a structure that will withstand extreme weather or flooding, then print and test their models.

LEARNING OBJECTIVES

After completing this project, students will be able to:

› Create objects in Tinkercad using the **group**, **hole**, and **transformation tools**.

› Understand basic types and patterns of weather.

› Understand conceptual methods of creating structures impervious to flooding.

VIRGINIA STANDARDS

SOL: SCI 2.6 The student will investigate and understand basic types, changes, and patterns of weather. Key concepts include

› Temperature, wind, precipitation, drought, flood, storms

› The uses and importance of measuring and recording weather data

NGSS STANDARDS

MS-ESS2-4 Earth's Systems Develop a model to describe the cycling of water through Earth's systems driven by energy from the sun and the force of gravity.

3-5-ETS1-1 Engineering Design Define a simple design problem reflecting a need or a want that includes specified criteria for success and constraints on materials, time, or cost.

3-5-ETS1-2 Engineering Design Generate and compare multiple possible solutions to a problem based on how well each is likely to meet the criteria and constraints of the problem.

3-5-ETS1-3 Engineering Design Plan and carry out fair tests in which variables are controlled and failure points are considered to identify aspects of a model or prototype that can be improved.

TEACHER PREPARATION

A. Create Tinkercad accounts: Students must have an active account in order to design their structures.

B. Place sand or stones at the bottom of the plastic bin to replicate soil and to create a firm base for testing the printed structures.

C. Use a small book to prop the bin up at a slight incline.

STUDENT ACTIVITY

STEP 01:
DISCUSS AND BRAINSTORM

A. Discuss how communities deal with weather challenges, such as floods and hurricanes.

B. Research how heavy rain can cause flooding; use this information to create a structure that will withstand a flood.

C. Brainstorm which type of structures would be most beneficial in a flood-prone environment.

STEP 02:
TINKERCAD INSTRUCTION

In order for you to successfully complete this project you will need a basic understanding of how to use Tinkercad software. Review and practice the following tools before building your structure.

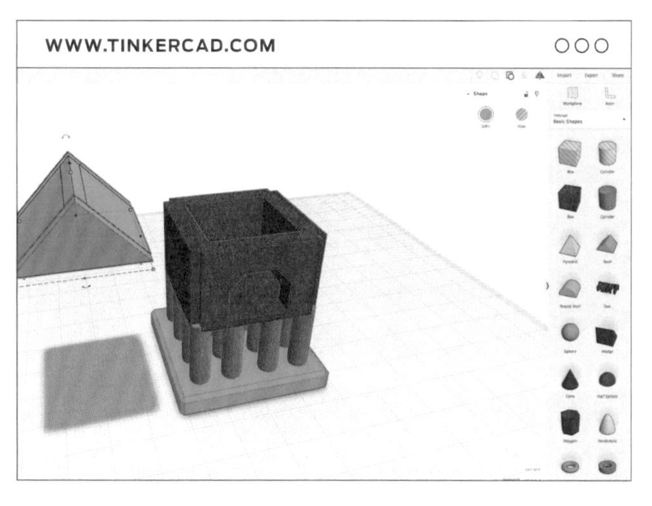

A. Move: This includes rotate, scale, etc.

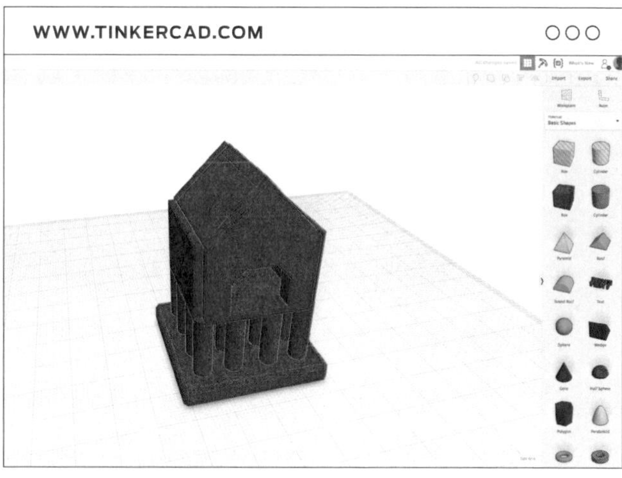

B. Group/Ungroup: These functions allow you to merge one or more shapes to create your weatherproofing structures. This tool also allows you to use holes to cut from an existing shape. Grouping is an essential step in exporting files for 3D printing.

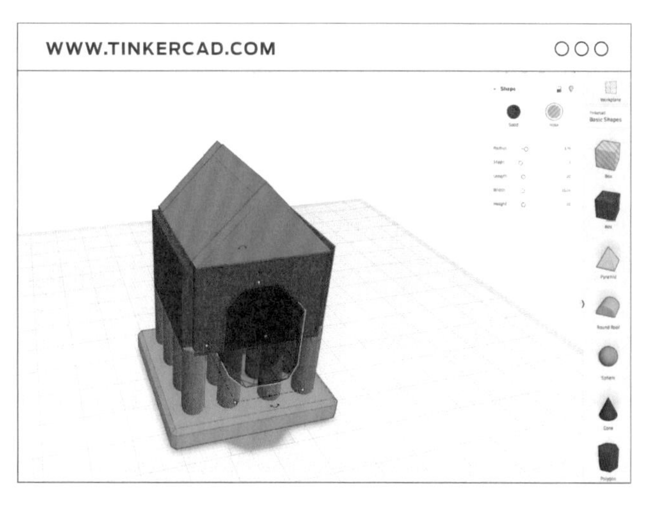

C. Hole: Holes in Tinkercad are shapes that are used to remove material from another object. You can use holes to create windows, doors or other cutouts.

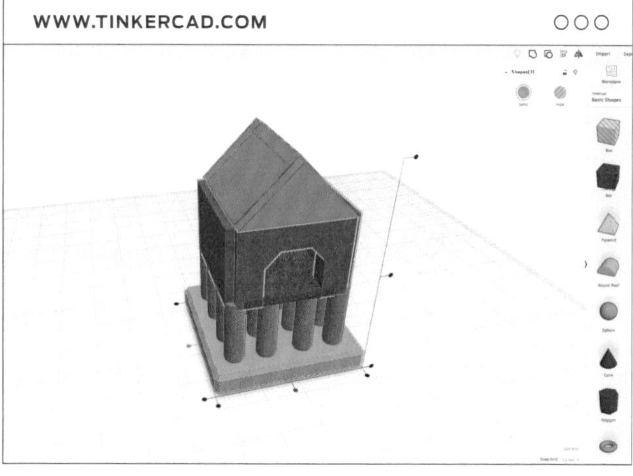

D. Align: Align can save you time when moving objects. This tool allows you to line up edges or centers of models.

STEP 03:
DESIGN

A. Design structures: Once you are familiar with the basic operation of Tinkercad software, start designing your own flood-proof structure. This should take 1 - 2 class periods for design and iteration.

B. Review structures: Once you have completed your design, review it with your group and teacher. Make sure all parts of your model are touching so that it will print successfully. Also make sure your model is grouped into one part.

STEP 04:
PRINT

A. Export: Click **export** in the upper right side of the Tinkercad software window. Select the .STL format to start the download.

B. Prepare: Once you have completed and exported your model, import it into MakerBot Print to prepare for printing.

C. Print all models with the same settings to ensure consistent testing.

Print Settings:

Rafts	Yes
Supports	Depends on Model
Resolution	0.2mm
Infill	5%

STEP 05:
TEST

One at a time, test each group's model in a mock flooding simulation to see how well each structure performs under heavy rain and flooding. Keep the following questions in mind:

› **Does the model stay in place?**

› **Are there parts of the model that could easily be washed away?**

› **Does the model stand above the water line?**

A. Place structure into the water bin, using the sand/ stones as your soil.

B. Add water to the bin to simulate heavy rain. Observe the behavior of the model once water reaches flood levels.

STEP 06:
REFLECT

A. Complete the worksheet: Answer the following questions on the provided worksheet:

› Did your structure stand up to the flood? Explain what happened when you tested it.

› If you had a chance to redesign, what would you do differently?

PROJECT WORKSHEET

WEATHER SURVIVAL CHALLENGE

STUDENT NAME

DRAW YOUR STRUCTURE:

Did your structure stand up to the flood?
Explain what happened when you tested it.

If you had a chance to design your structure, what would you do differently?

TO DOWNLOAD: *thingiverse.com/thing:1711704*

PROJECT COMPLETE:
REIMAGINE AND REBUILD

GOING FURTHER

A. Have students redesign their structures using what they learned from their first test. Test the new models and compare the results.

B. Change some of the print settings, such as increasing the infill density, and run the same flood test again: what happens?

C. Now that students have designed structures for a flood environment, ask how they would design structures to withstand other severe weather like tornados, hurricanes, etc.

PROJECT EIGHT

RUBBER BAND GLIDERS

LINK: *thingiverse.com/thing:1744950* for access to handouts, videos and other materials associated with this project.

PROJECT INFO

AUTHOR
Jeffry Turnmire,
@*insane66*

SUBJECT
Engineering

AUDIENCE
Grade Levels 6–12

DIFFICULTY
Intermediate

SKILLS NEEDED
Basic Autodesk®
Fusion 360™ experience

DURATION
3–5 Class periods

GROUPS
15 Groups
1–2 students / Group

MATERIALS

Rubber Bands

Tape Measure

Tape or Superglue

Calipers or Ruler

Open space to launch

SOFTWARE
Autodesk Fusion 360

PRINTERS
Works with all MakerBot® Replicator® 3D printers, except the MakerBot Replicator Mini

PRINT TIME
Prep: 1 hr / Glider
Lesson: 30 mins / Group

FILAMENT USED
½ Large Spool

"I'm a pilot and have loved airplanes since I was old enough to talk. I have built hundreds of model planes. Some fly, some just look pretty. We need more young people interested in aerospace! So, I set out to design a project to make a practical exploration of wing design by making rubber band gliders with wings that can easily be replaced, redesigned, and tested."

– **Jeffry Turnmire**

LESSON SUMMARY

Rubber band gliders are a great way to get students excited about engineering, specifically flight and aerodynamics.

In this project you'll provide students with a universal rubber band glider body to be paired with wings that they design in class. With a large selection of wings available on Thingiverse, as well as the wings designed by your class, students will learn basic principles of aerodynamics through practical research and exploration as the gliders crash and soar.

It's also a great excuse to spend a fun class period with your students launching rubber band gliders!

LEARNING OBJECTIVES

After completing this project, students will be able to:

- **Learn** about aerodynamic forces including lift, drag, weight, thrust, and flight
- **Describe** how wing design influences lift
- **Collect data** and modify wing design based on this data

COMMON CORE MATH STANDARDS

RST.11-12.8 Science & Technical Subjects Evaluate the hypotheses, data, analysis, and conclusions in a science or technical text, verifying the data when possible and corroborating or challenging conclusions with other sources of information.

CCSS.Math.Practice.MP4 Model with Mathematics By high school, a student might use geometry to solve a design problem or use a function to describe how one quantity of interest depends on another.

NGSS STANDARDS

HS-ETS1-2 Engineering Design Design a solution to a complex real-world problem by breaking it down into smaller, more manageable problems that can be solved through engineering.

HS-ETS1-3 Engineering Design Evaluate a solution to a complex real-world problem based on prioritized criteria and trade-offs that account for a range of constraints, including cost, safety, reliability, and aesthetics as well as possible social, cultural, and environmental impacts.

TEACHER PREPARATION

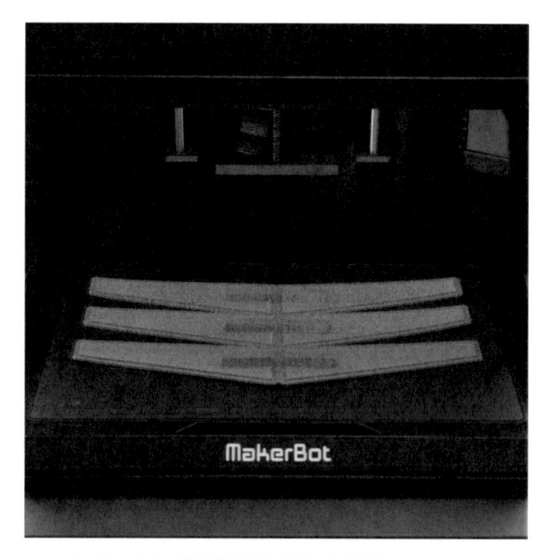

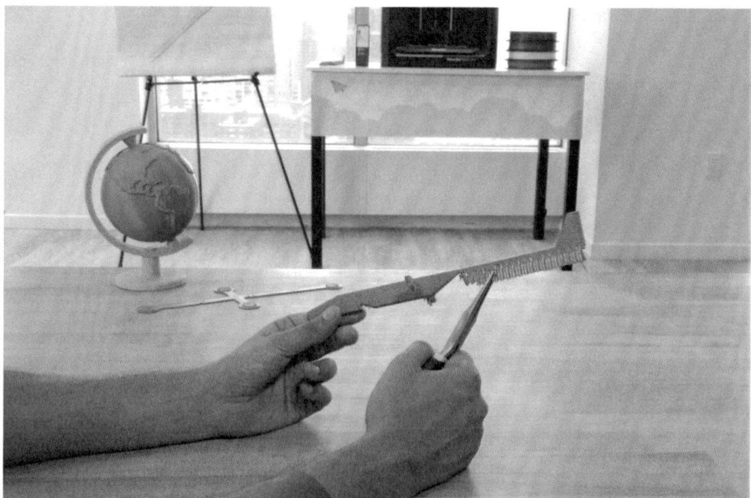

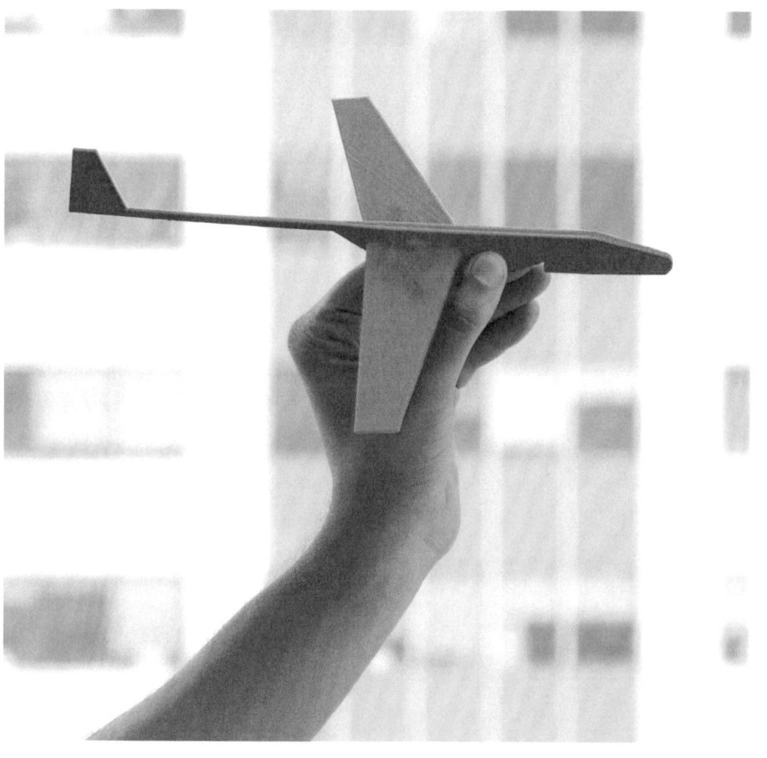

A. Print two or three of the glider body files included in the Thingiverse Education post (*thingiverse.com/thing:1744950*).

TIP: When printing the glider body, print with support material and rafts. Printing the wings laying flat on the build plate creates a higher-quality wing.

B. Print a few of the sample wings to give students ideas about different wing designs.

C. Review and discuss some of the basic principles and terminology of aerospace engineering:

› **Aerodynamic force:** The force exerted on a body whenever there is relative velocity between the body and the air.

› **Friction**: The resistance that one surface or object encounters when moving over another.

› **Air resistance:** The friction force air exerts against a moving object.

› **Center of gravity**: The point at which the entire weight of a body may be considered as concentrated so that if supported at this point the body would remain in equilibrium in any position.

D. Discuss:

› How air resistance, friction, and aerodynamic forces are considered in wing design, NASA has great online resources to help

› How center of gravity relates to glider performance

› Different styles of wings like rectangular, elliptical, swept (forward and rear), and delta

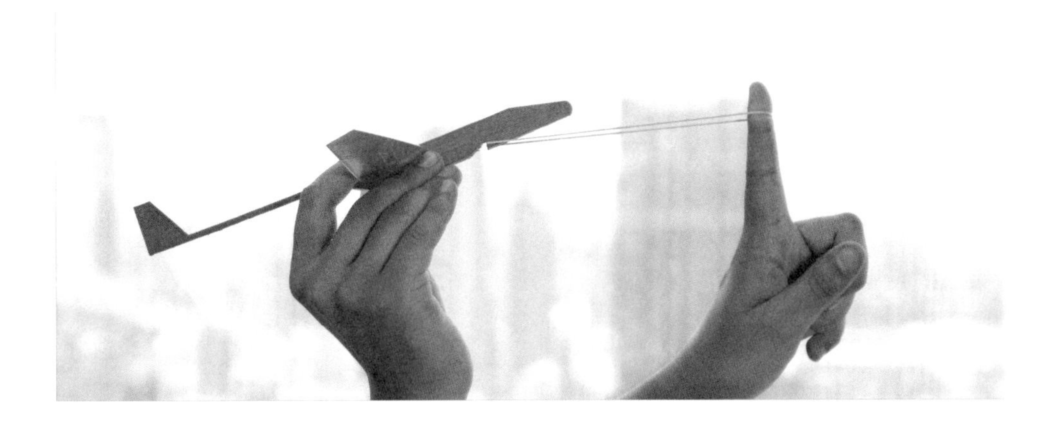

E. Demonstrate:

› 3D print and launch one of the predesigned gliders.

› Discuss which wing designs are more and less effective.

STUDENT ACTIVITY

STEP 01:
DESIGN AND BUILD

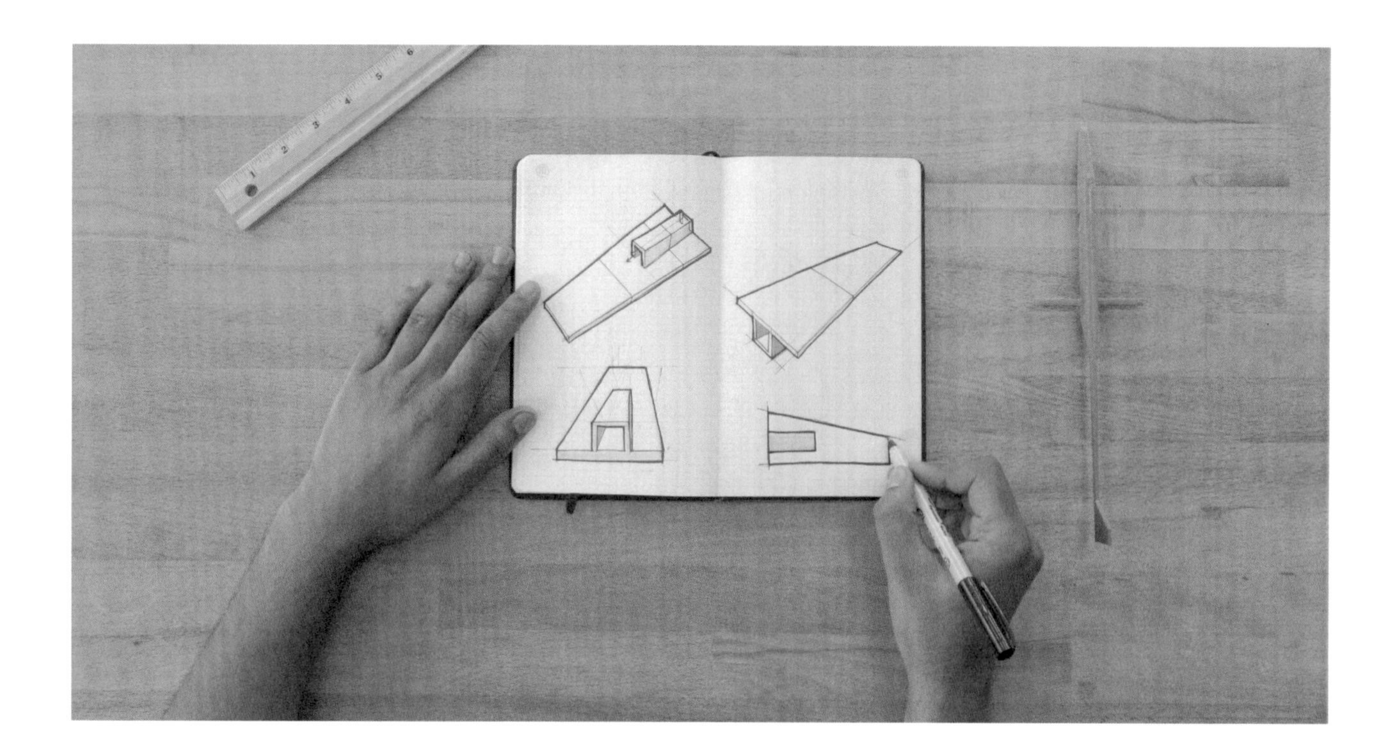

A. Begin planning a set of glider wings using the knowledge you gained through classroom discussion and demonstration. Your understanding of aerodynamic forces and flight will be tested later.

B. Open Autodesk Fusion 360 and design a set of wings to attach to the pre-made glider body. It is useful to have the printed gliders available to hold and measure as you design.

TIP: The original Fusion 360™ software files are included in the Thingiverse Education page for this project. If needed, students can download and open those to help make sure their wing designs will fit on the body.

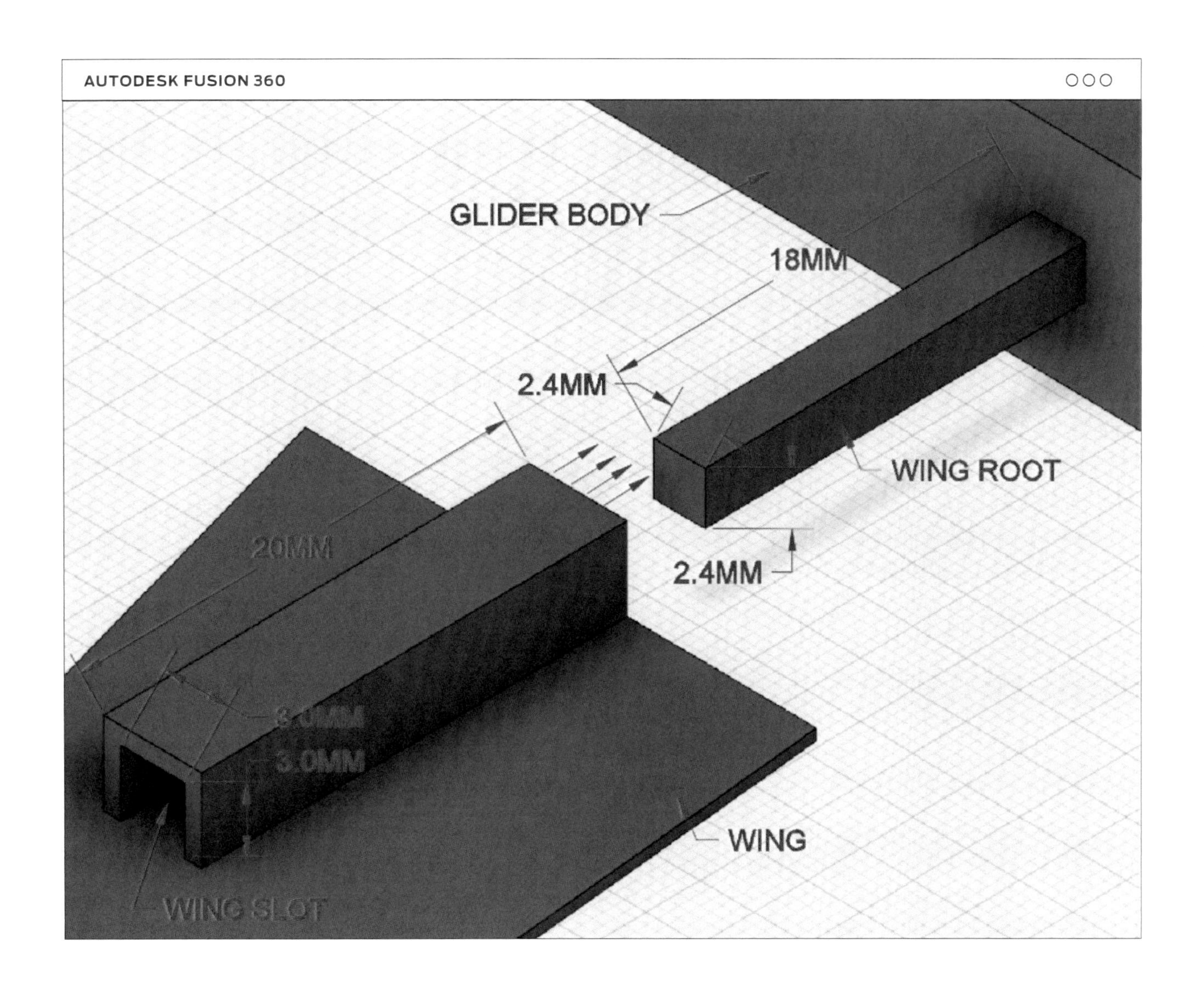

C. Review the diagram: The wing root on the glider body is 2.4 mm square x 18mm long. Your wings will need a socket that measures roughly 3 mm square x ~20mm long to slip onto the root.

TIP: Wing weight is one of the most important factors affecting flight. Experiment with the weight of your design by changing print settings like infill.

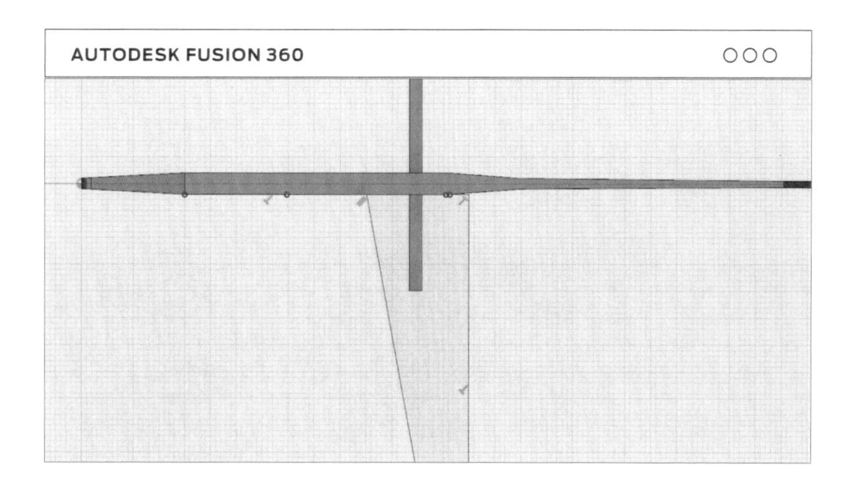

D. Sketch / extrude wing: Make sure your wing has enough space to fit the wing root.

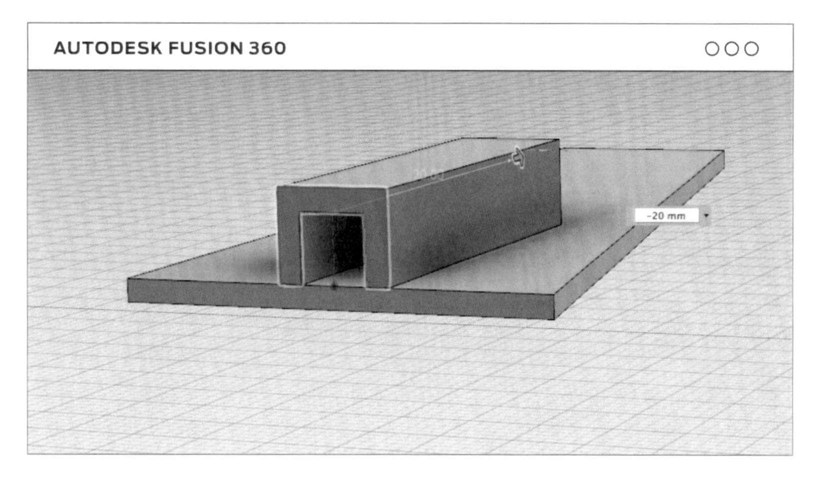

E. Sketch / extrude the root: Make sure it is connected to your wing and has the correct dimensions.

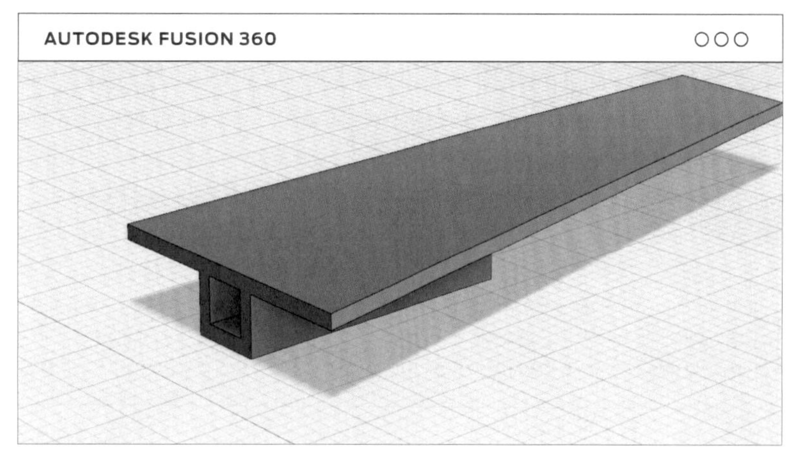

F. Use the mirror tool to create a second wing. **Move** and **rotate** the wings to position them on the glider body and verify the size and fit.

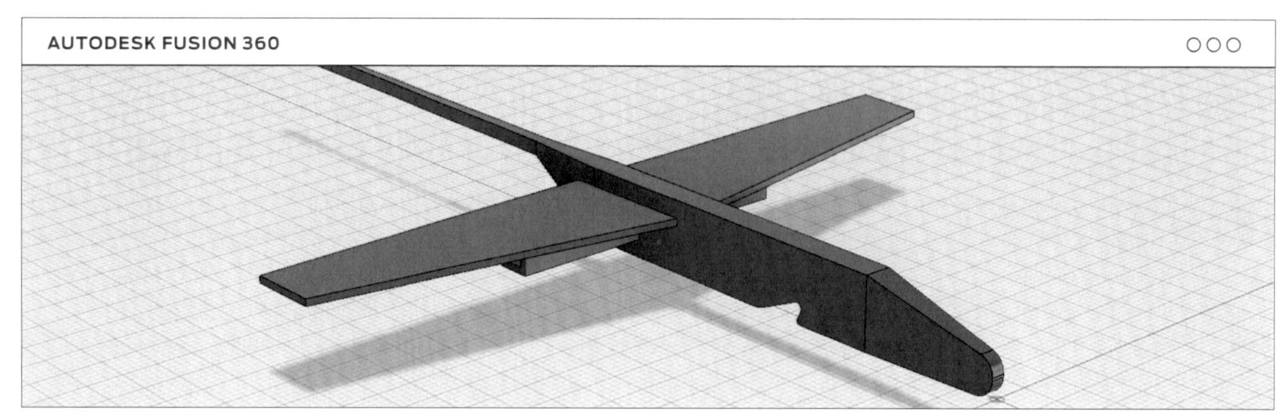

STEP 02:
PRINT

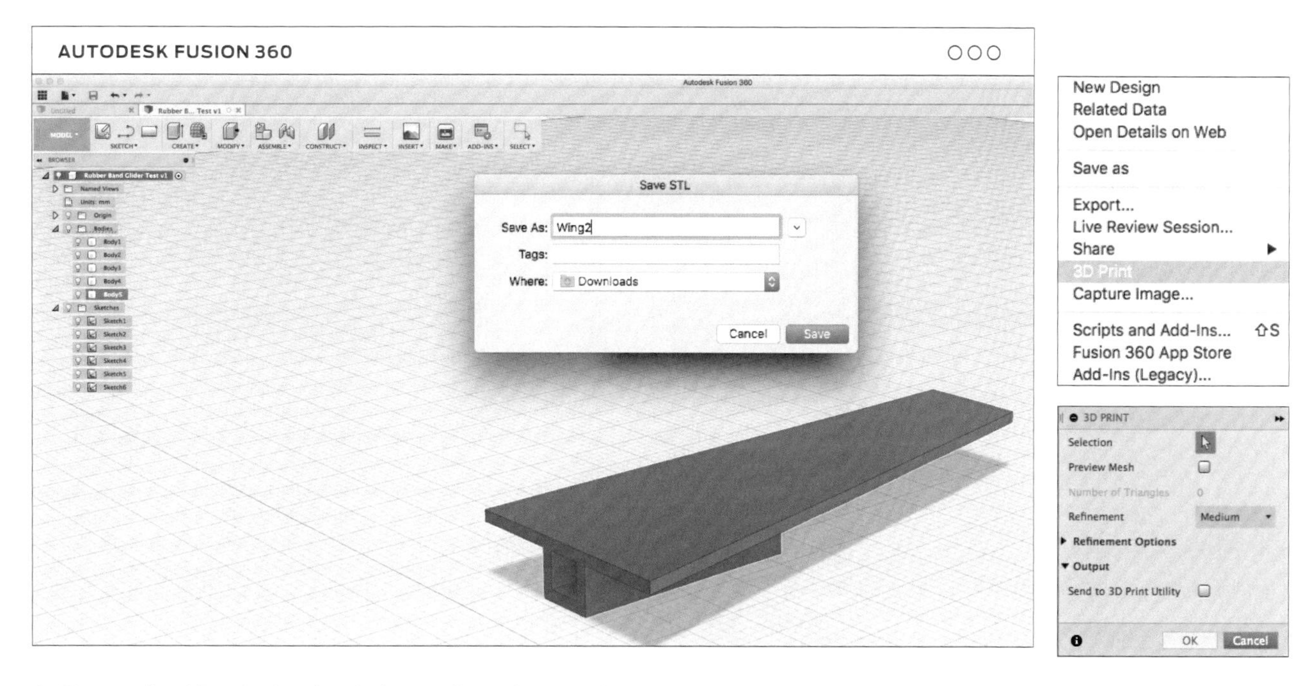

A. Export for 3D printing by clicking **File,** selecting the **3D Print menu** and choosing the **.STL** format.

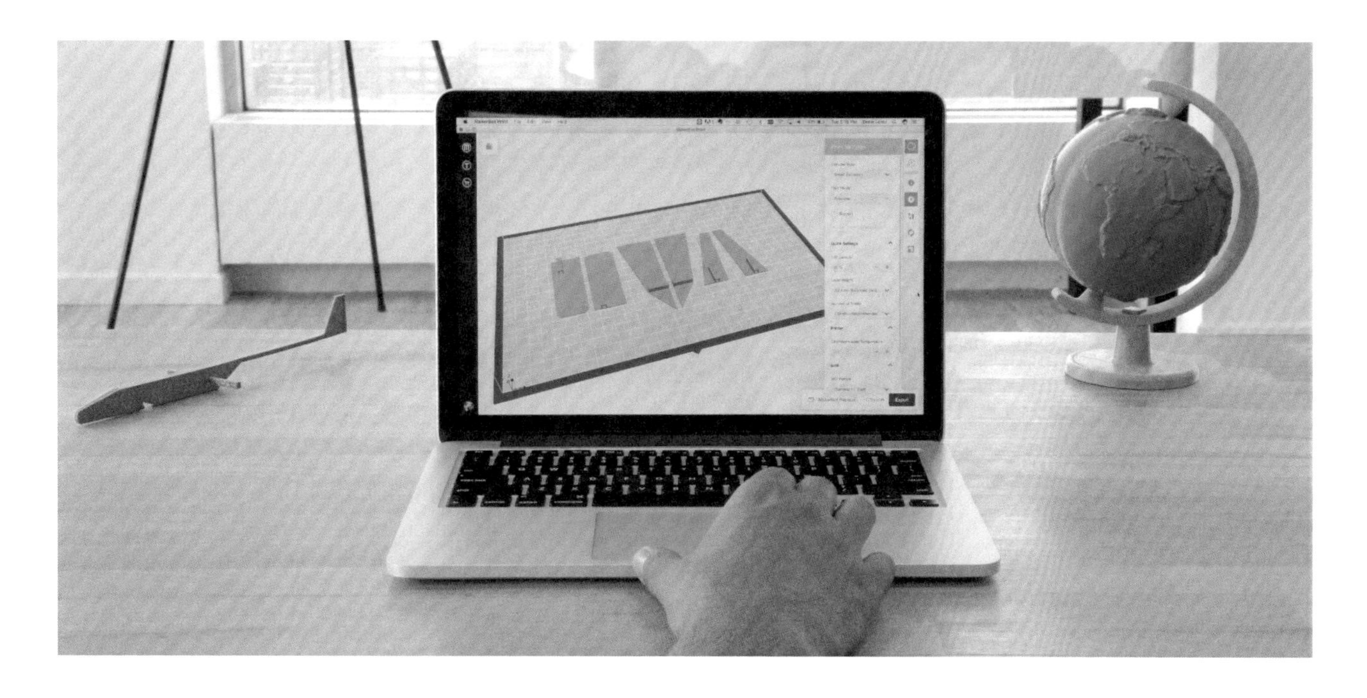

B. Import the .STL files into MakerBot Print and prepare for printing.

Print Settings:

Rafts	Yes
Supports	Yes (body only)
Resolution	0.2mm
Infill	30%

TIP: Have students experiment with different iterations of the glider's body to perfect its center of gravity.

STEP 03:
TEST

A. Assemble the gliders by slipping the wings onto the body roots. Set up your testing range with a launch line (marked with tape) and a large open space. If possible, lay out a landing grid to make measuring easier after launching.

CAUTION: Be aware of what is down range. With stronger rubber bands or different wind conditions, gliders could fly 30' or more!

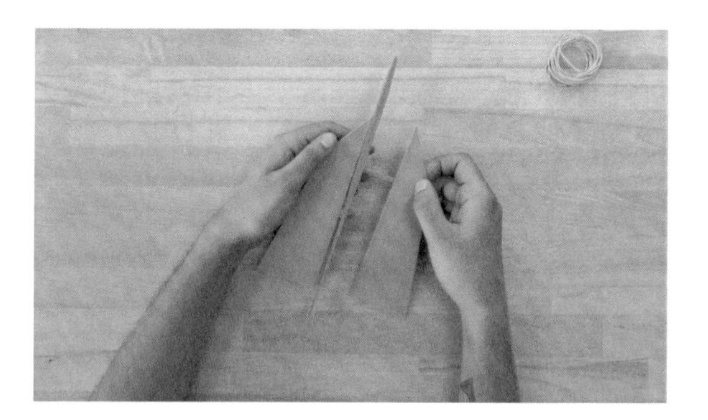

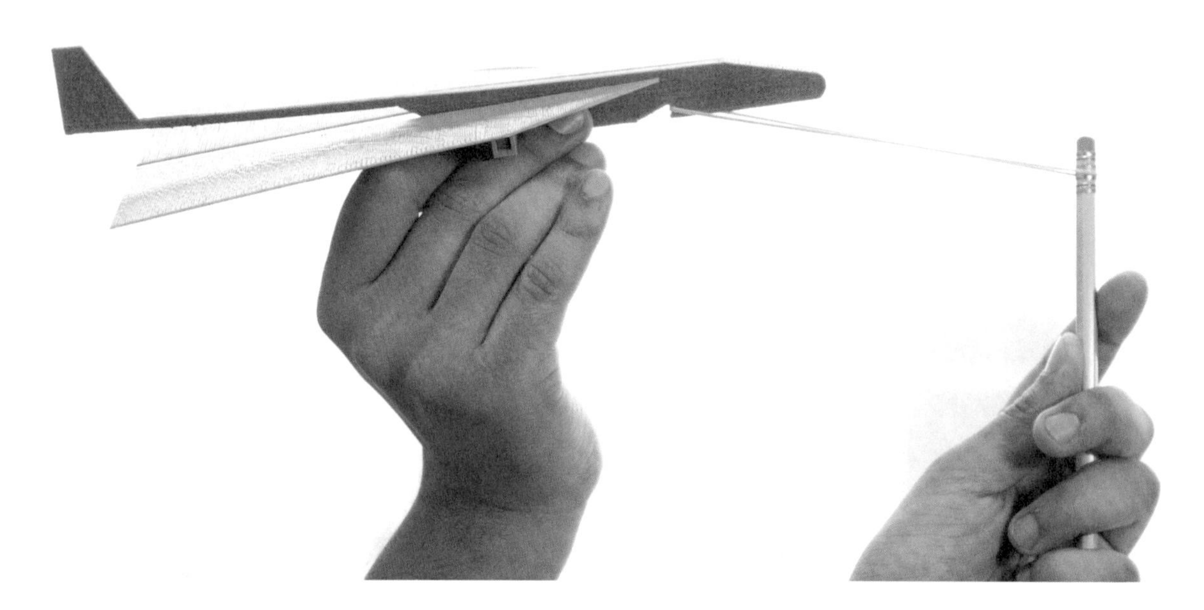

B. Launch gliders from the launch line one at a time using the same rubber band. If you find the gliders hitting your fingers, use a pen as a launching tool. It's important that everyone uses the same method for consistency.

TIP: Launches should be performed at the same angle, into similar wind conditions. Watch the video on Thingiverse demonstrating proper launching technique.

C. Measure the flight distance of each glider and record all data in a table.

STEP 04:
EVALUATE

A. Evaluate your data for trends and calculate the average distance that you were able to achieve with your wing design.

STEP 05:
REVIEW

A. Discuss with your class:

› Which wing designs were most effective?

› Do they share common design traits?

› What other conclusions can you draw from the class results?

B. How could you do the experiment differently?

› What outside factors affected the results and how can they be minimized in the future?

PROJECT WORKSHEET

RUBBER BAND GLIDERS

STUDENT / GROUP NAME	FIRST LAUNCH	SECOND LAUNCH	AVERAGE DISTANCE

TO DOWNLOAD: thingiverse.com/thing:1744950

PROJECT COMPLETE:
LEARN TO FLY

GOING FURTHER

A. After the first round of testing you can prompt students to use the iterative design process and design a new set of wings. In future class periods, launch student gliders again and test their improved designs compared to their original designs.

B. Have students start from scratch and design both the glider bodies and wings. This is more time-intensive but allows for more design creativity.

PROJECT NINE

ARCHIMEDES SCREW BONANZA

LINK: *thingiverse.com/thing:1769714* for access to handouts, videos and other materials associated with this project.

PROJECT INFO

AUTHOR
MakerBot Learning
@makerbotlearning

SUBJECT
Engineering

AUDIENCE
Grade Levels 9–12

DIFFICULTY
Advanced

SKILLS NEEDED
Some Onshape®
software experience

DURATION
10—12 Class Periods

GROUPS
3–4 Groups
6–7 Students / Group

MATERIALS
2 Large Bowls
Superglue
Bag of Dried Beans
Dual Lock Tape
Ruler or Calipers

SOFTWARE
Onshape (**web app**)

APPLICABLE PRINTERS
Works with all
MakerBot® Replicator®
3D printers

PRINT TIME
Prep: 65 hrs
Lesson: 25—30 hrs / Group

FILAMENT USED
3–4 Large Spools

"This projects brings an ancient invention to life using modern technology. While the project is definitely an investment in printing time and filament, the students' final products make this lesson well worth the effort!"

– **MakerBot Learning**

LESSON SUMMARY

The Archimedes Screw is a device believed to be invented by Archimedes, one of the world's greatest scientists, in the 3rd century B.C. It was most often used to transport water from lower ground to higher ground for irrigation, flood prevention, and more. Interestingly, it is still used today in many different applications.

Well before power tools were created, people had to cleverly use common materials to make their work more efficient. The Archimedes screw is one example of this type of simple machine. A basic design consists of a screw (helical spiral part), a tube (full or partial), and a crank. By rotating the crank, you can move large quantities of material up an incline much more efficiently than by simply carrying it. All Archimedes screw designs have the same basic components, but can take many shapes and sizes depending on the application. Check out some videos in the Thingiverse Education post (thingiverse.com/thing:1769714).

In this project, students will create or modify an Archimedes screw to transport material from lower ground to higher ground. The sample file included was designed in Onshape. The students' task is to observe the sample printed screw design, and create a more efficient screw design that fits into the supplied case.

LEARNING OBJECTIVES

After completing this project, students will be able to:

› **Discuss** the history of the Archimedes screw

› **Apply** engineering principles to design a simple machine

› **Comprehend** and apply the fundamentals of parametric 3D design

› **Understand** the impact of changing design parameters

NGSS STANDARDS

HS-PS3-3: Energy Design, build, and refine a device that works within given constraints to convert one form of energy into another form of energy.

HS-ETS1-2: Engineering Design Design a solution to a complex real-world problem by breaking it down into smaller, more manageable problems that can be solved through engineering.

HS-ETS1-4: Engineering Design Use a computer simulation to model the impact of proposed solutions to a complex real-world problem with numerous criteria and constraints on interactions within and between systems relevant to the problem.

TEACHER PREPARATION

CASE BOTTOM

CRANK

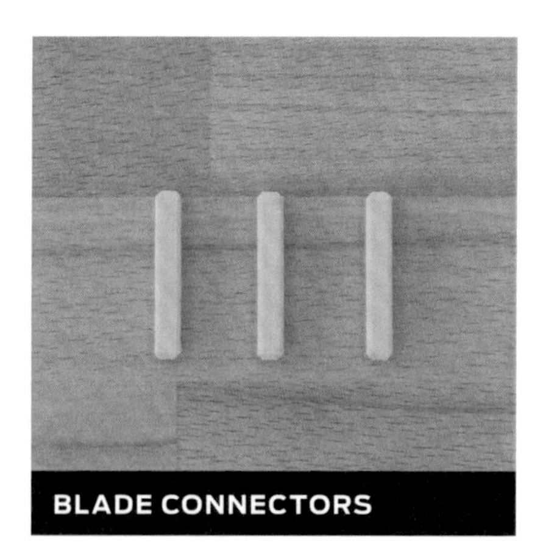
BLADE CONNECTORS

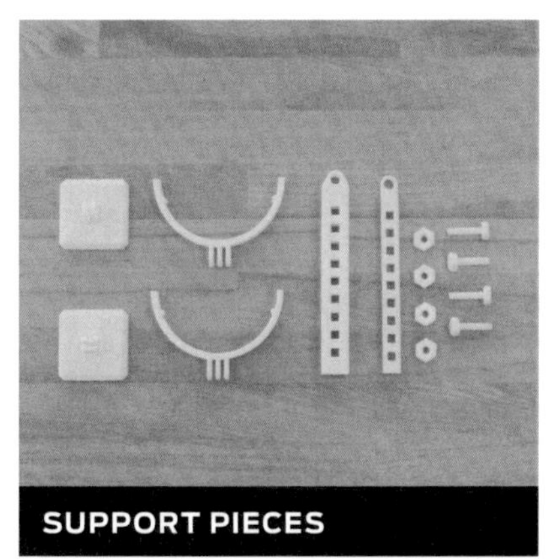
SUPPORT PIECES

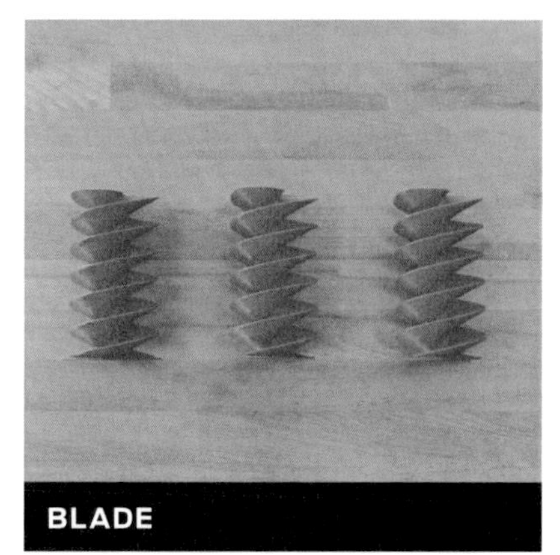
BLADE

SPILL PLATE

BOTTOM ADAPTER

CASE BRACKET

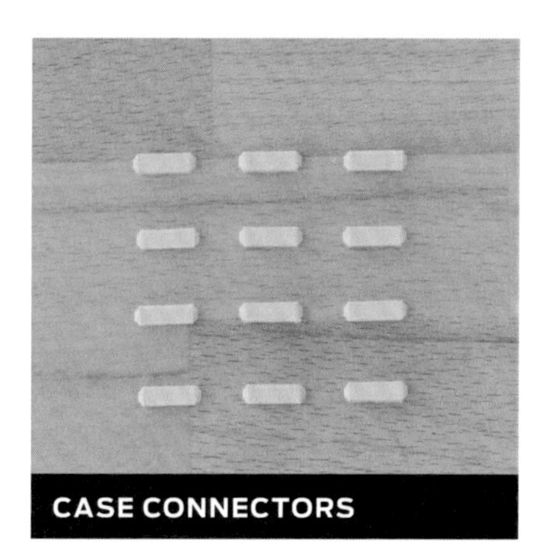
CASE CONNECTORS

A. 3D print the sample files from Thingiverse: Leave yourself at least a week for printing and assembly. Print the following quantities of parts:

- 3x Case Bottom
- 3x Blade
- 1x Bottom Adaptor
- 2x Case Bracket
- 1x Crank
- 2x Support Clamp
- 1x Support Bar Top
- 1x Support Bar Bottom
- 2x Support Base
- 4x Support Nut and Bolt (you can subsitute 1/4-20 Nut & Bolt)
- 12x Case Connector (24x if you print the Case Top parts)
- 4x Blade Connector
- 1x Spill Plate
- 3x Case Top (optional - only if you want to fully enclose the screw)

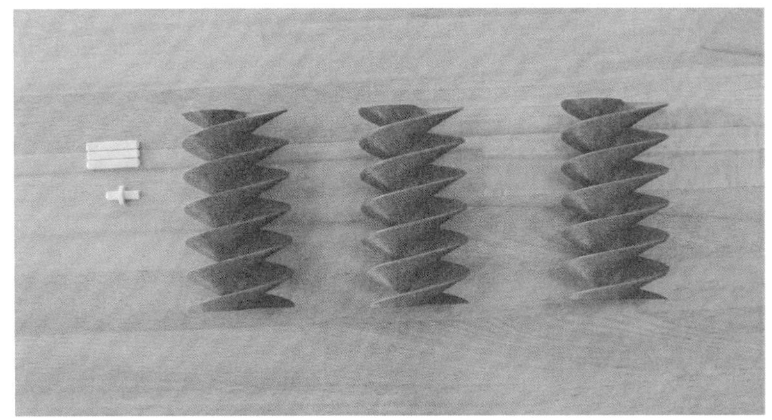

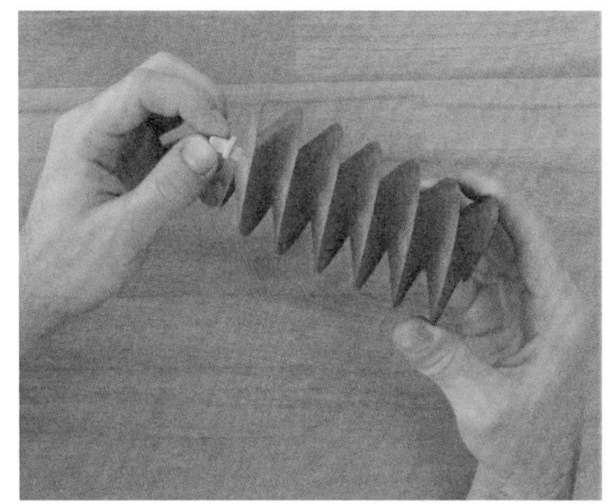

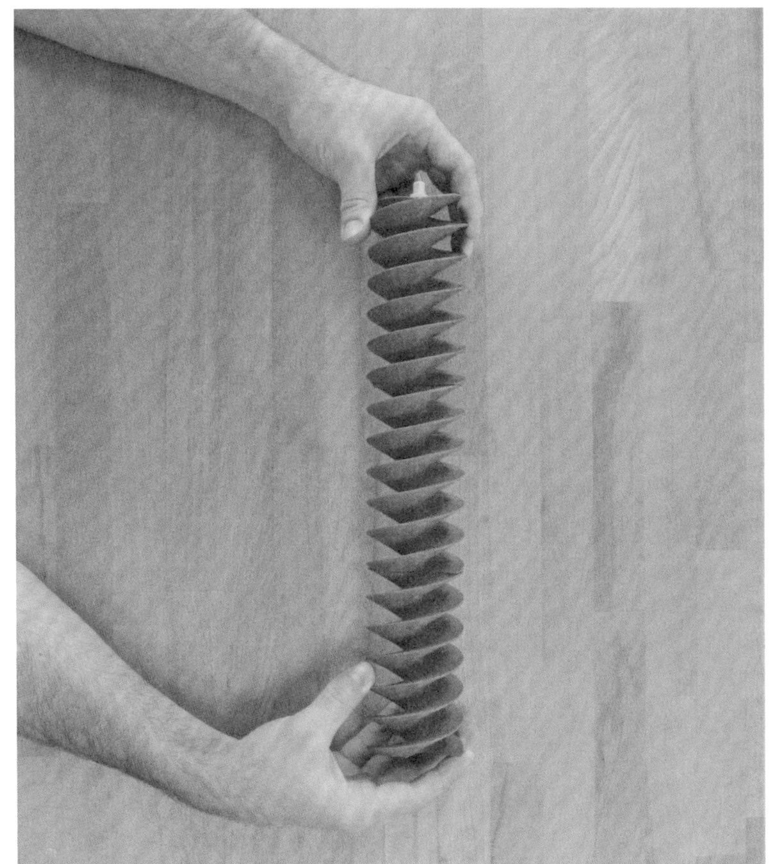

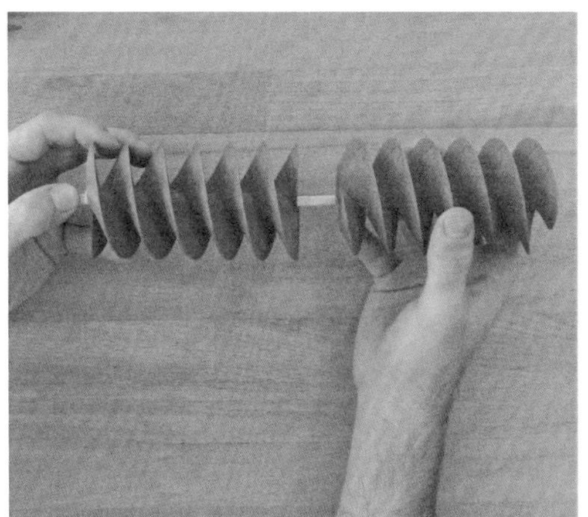

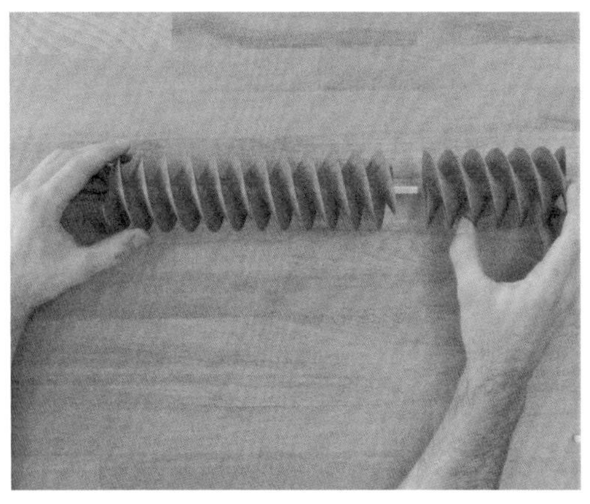

Print Settings:
Support Parts

Rafts	Yes
Supports	No
Resolution	0.2mm
Infill	25%

Print Settings:
All Other Parts

Rafts	Yes
Supports	No
Resolution	0.2mm
Infill	5%

B. Assemble the sample screw:
Glue a **blade connector** into the top hole of each **blade** part (3x). Once dry, connect the blade parts together and glue them into place.

Insert the square end of the **bottom adaptor** into the bottom hole of the blade assembly and glue into place.

TEACHER PREPARATION

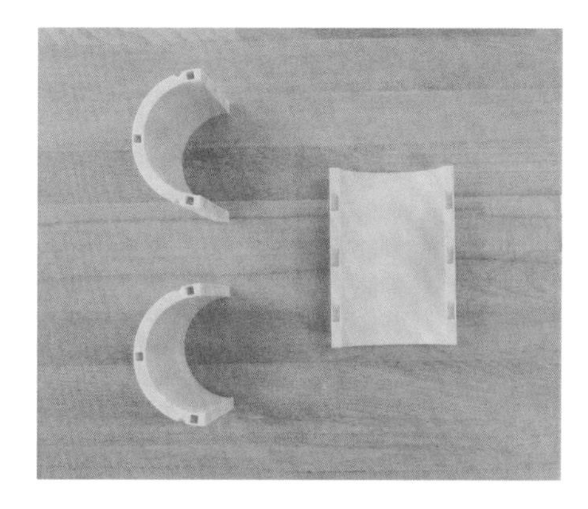

C. Assemble the case: Insert and glue **case connectors** into the 3 holes in the bottom of each **case bottom** part. Glue an additional 3 into the topmost **case bottom** part.

- Once dry, glue the 3 **case bottom** pieces together.
- Align a **case bracket** piece onto the **case connectors** at the bottom of the **case bottom** assembly. The pointed tip should be facing away from the **case bottom**. Glue the **case bracket** into place.

- Align the **spill plate** onto the center **case connector** at the top of the **case bottom** assembly. Glue into place.
- Insert the **case bracket** onto the **case connectors** at the top of the case, but **DO NOT** glue into place - this will allow you to remove the screw after demonstrating and replace it with student models.

Note: DO NOT glue the top bracket or crank - this will allow you to remove the screw after demonstrating and replace it with students' models later on. This way, you don't have to print a case for every group.

D. Assemble the supports:

› Insert the **support bar bottom** into a **support base** and line up the holes. Insert a **nut** and **bolt** to keep the parts together.

› Insert the **support bar top** into a **case clamp** and align the holes. Insert a **nut** and **bolt** to keep the parts together.

› Insert a **case clamp** into a **support base** and line up the holes. Insert a **nut** and **bolt** to keep the parts together.

› Attach the **support base** parts to the bowls using dual lock.

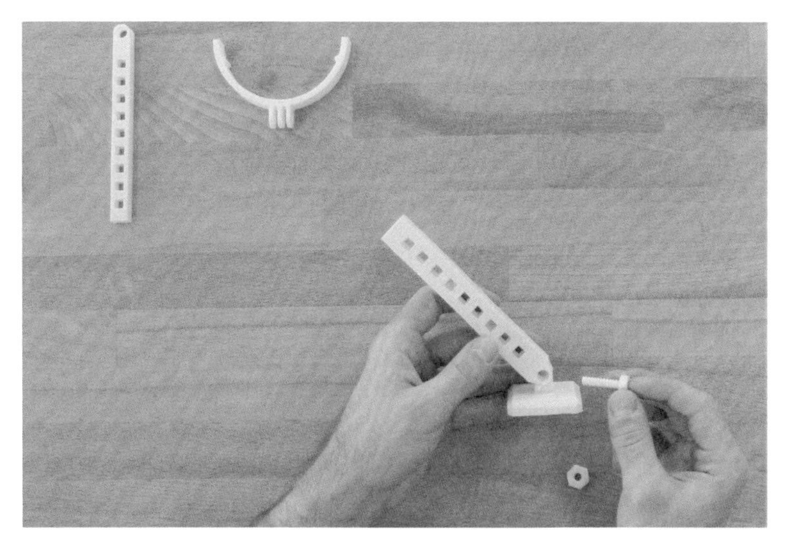

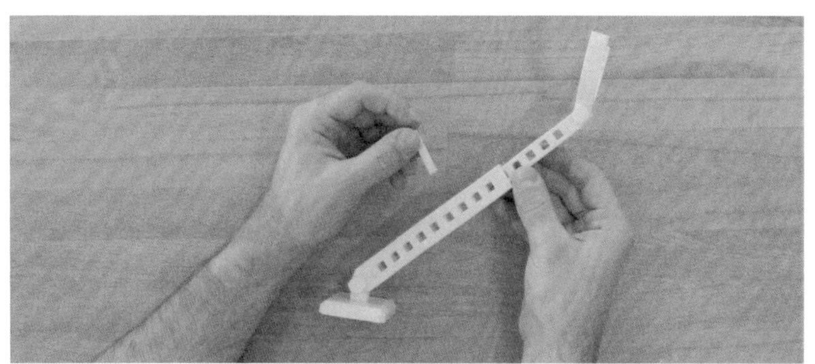

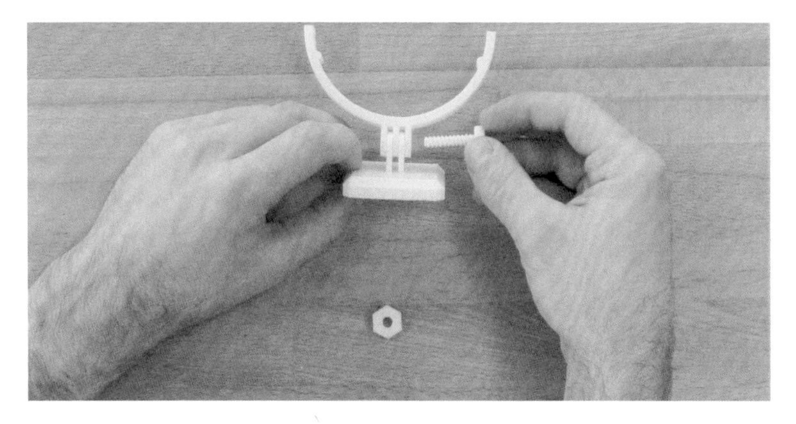

E. Final Assembly:

› Snap the case assembly onto the **case clamps**.

› Insert the blade assembly into the case by inserting the **bottom adaptor** into the **bottom case bracket**.

› Place a **case bracket** onto the top of the case by slipping it over the protruding **blade connector** and aligning it to the **case connectors**.

› FInally, slip the **crank** onto the **blade connector** and fill the bottom bowl with your material.

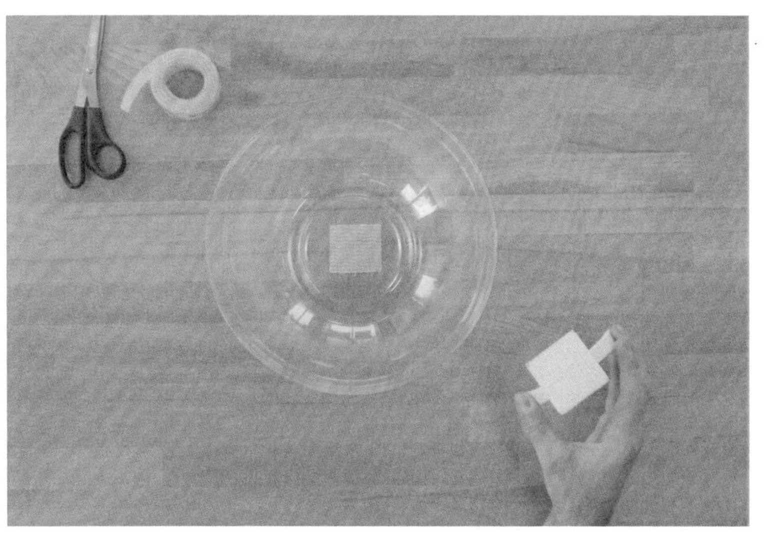

TEACHER PREPARATION

F. Demonstrate the sample screw design to students. Explain that their task is to make a more efficient screw design that fits into the case.

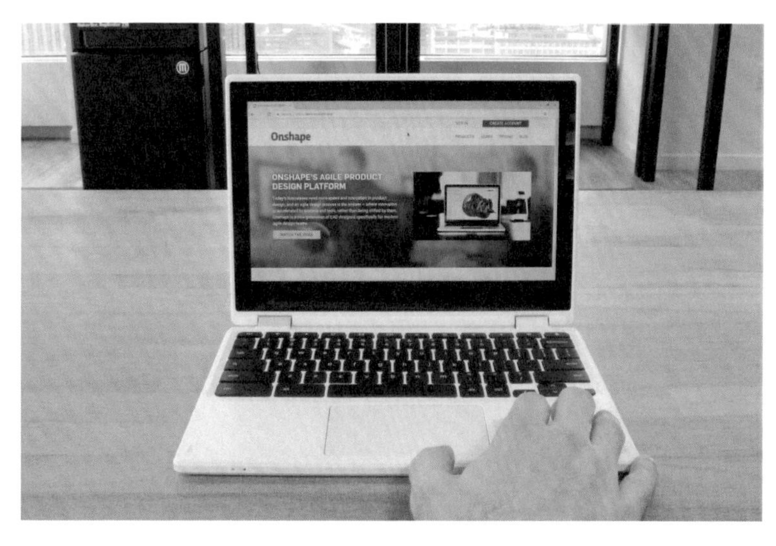

G. Create an Onshape account.

H. Distribute the sample Onshape file to students. The link to the file is in the Thingiverse Education post (*thingiverse.com/thing:1769714*).

STUDENT ACTIVITY

STEP 01:
PLANNING

A. Measure the bowls, case, transport distance, and material before beginning to plan your screw design.

B. Analyze the printed sample screw and take notes of possible improvements.

STEP 02:

REVIEW SAMPLE CAD DESIGN

In this step, you'll experiment with changing the variables in each feature of the sample CAD model and observe the impact.

A. Open your browser and navigate to the Onshape link included in the Thingiverse Education post (*thingiverse.com/thing:1769714*).

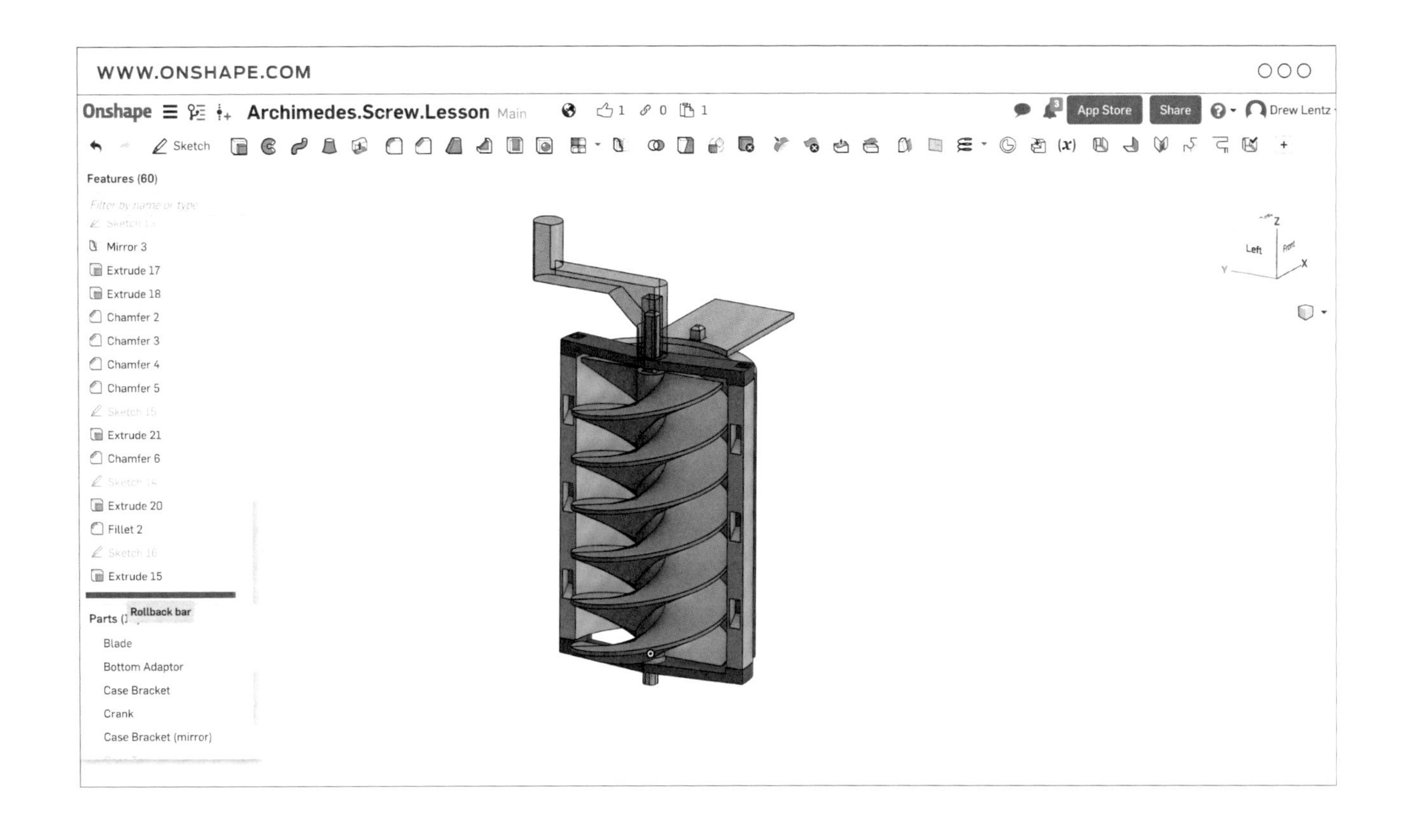

B. Review the major design steps for each screw part. The sample is by no means the only way to create this model—but should provide an idea for how to construct your own.

The basic steps to modeling this sample are:

› **Sketch** and **extrude** the center cylinder.

› Use **sweep** tool to create blades.

› **Sketch** and **extrude** case bottom, offset from blade. **Mirror** to create case top.

› **Sketch** and **extrude** the adaptors, brackets, crank, and spill plate.

TIP: Use the **rollback bar** in Onshape to go through the timeline and simulate the model being built.

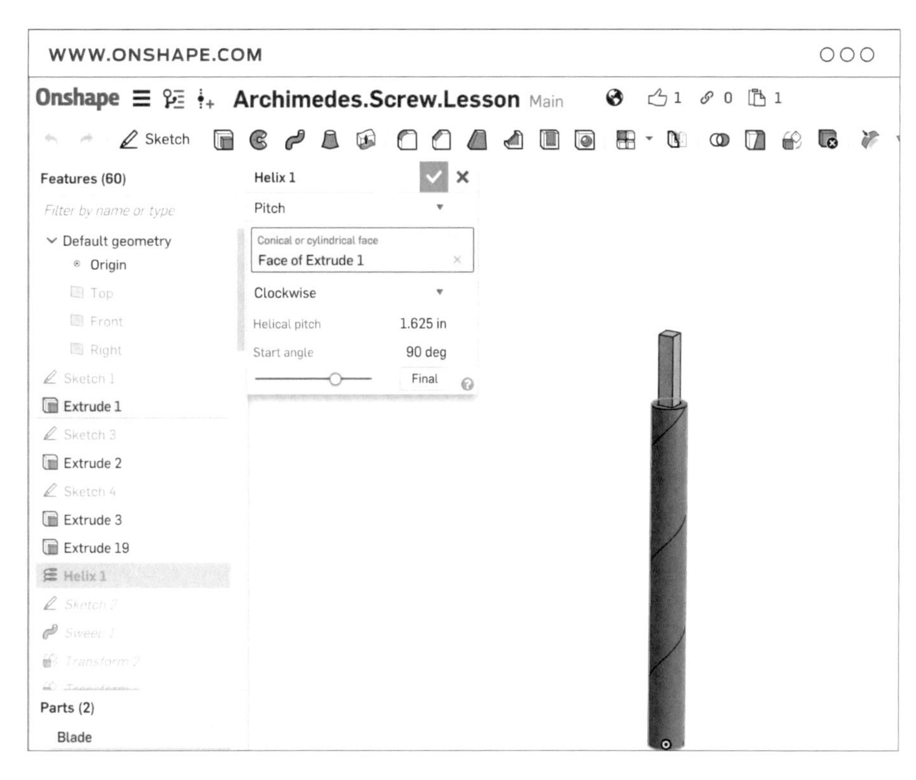

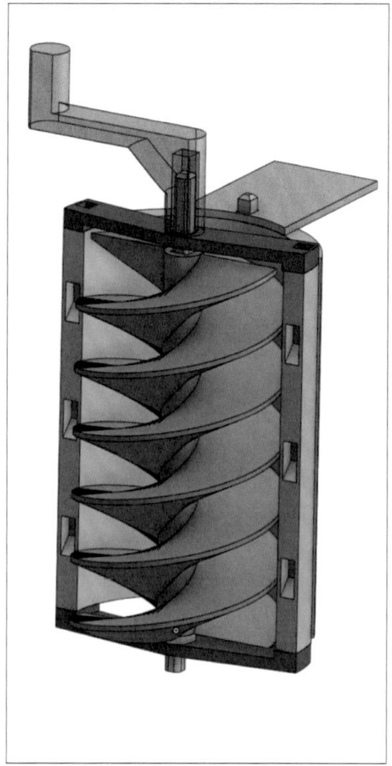

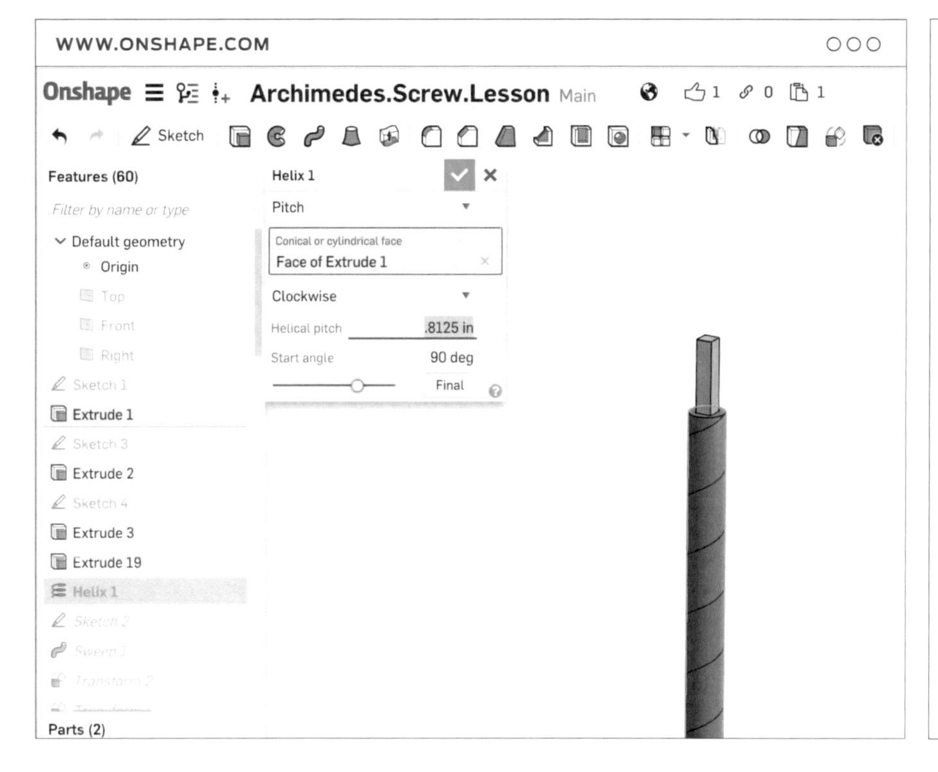

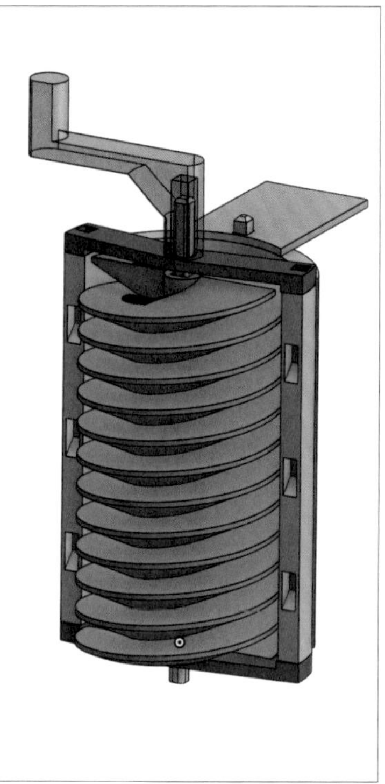

C. Edit the helix sketch: The **helix** sketch defines the pitch of the blade part. Right click to edit and experiment with this number to observe the impact on the blade. This is a quick way to dramatically affect the screw design. In these pictures, the pitch is reduced to create a tighter screw. Later, you will need to calculate an appropriate pitch for your blade.

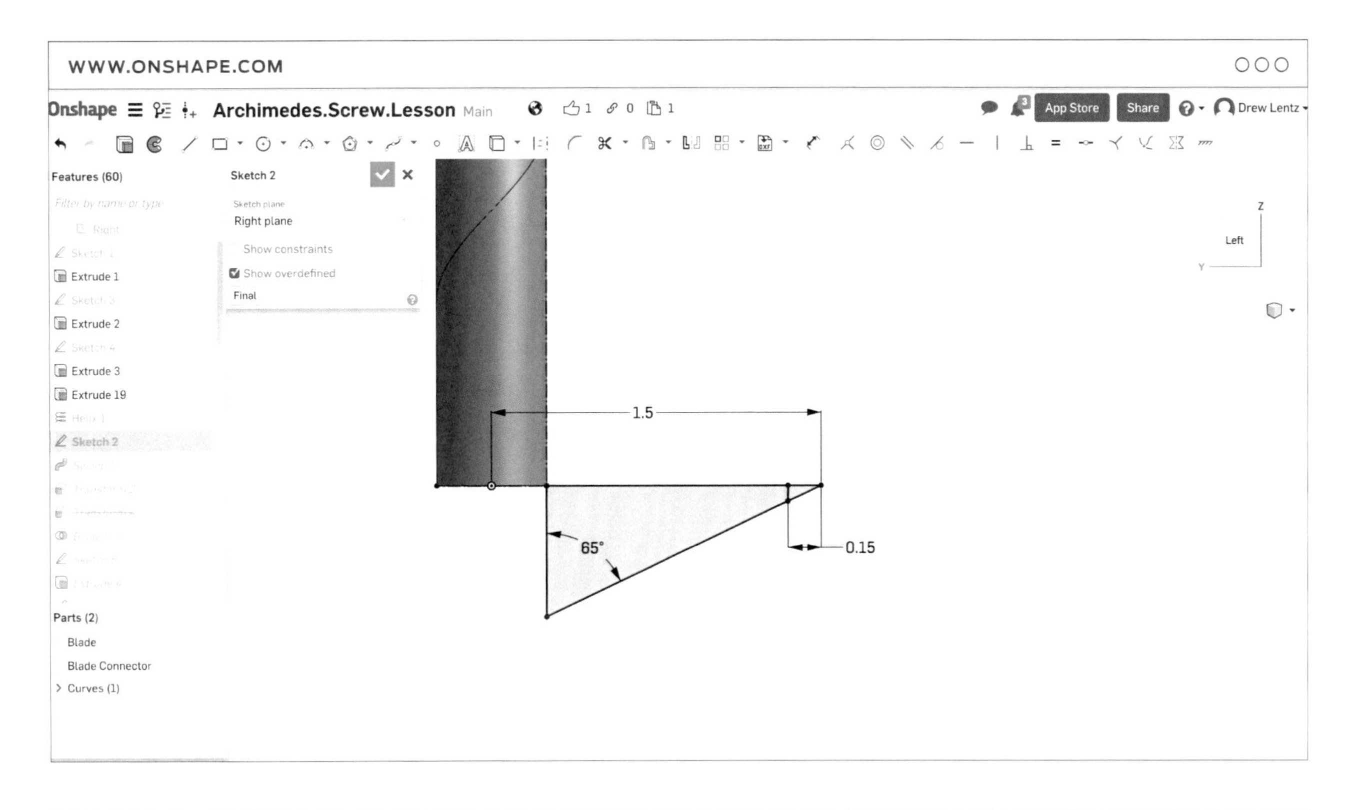

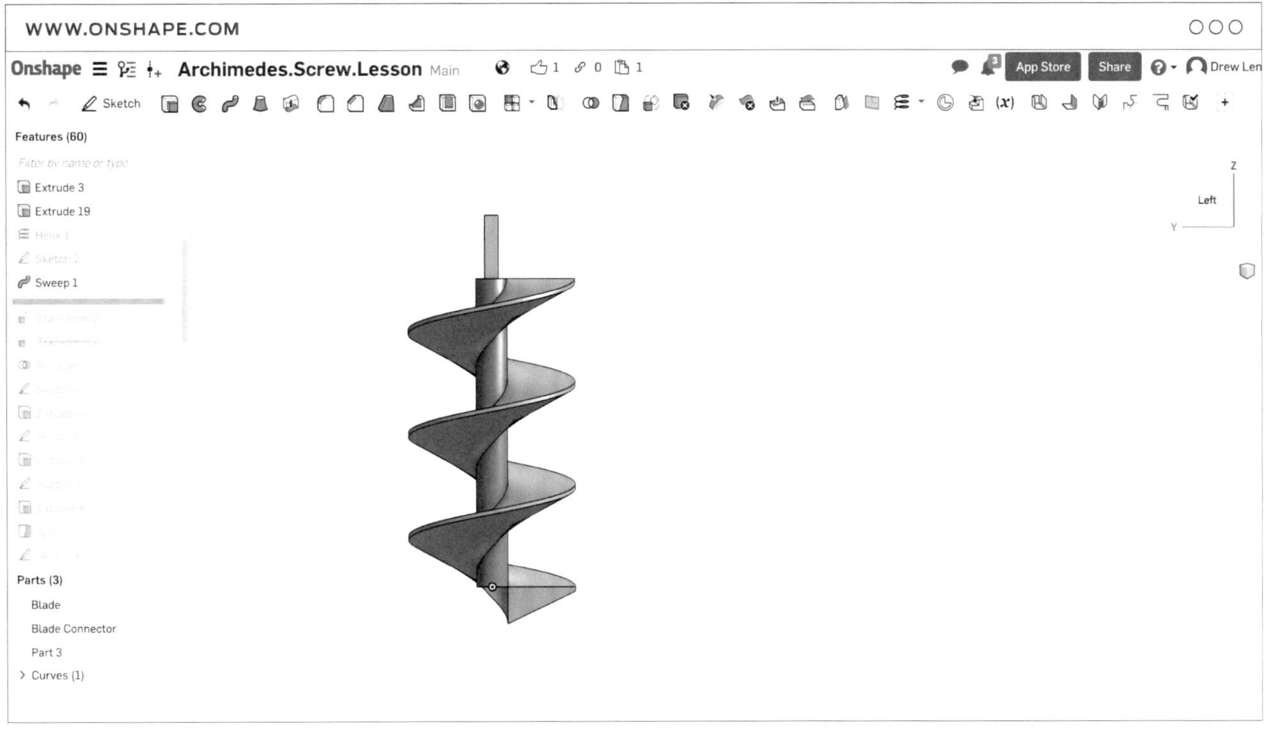

D. Edit the blade profile sketch: Experiment with changing the angle and length of the blade to observe the impact on the blade part. The 65-degree angle allows for 3D printing without the need for support material.

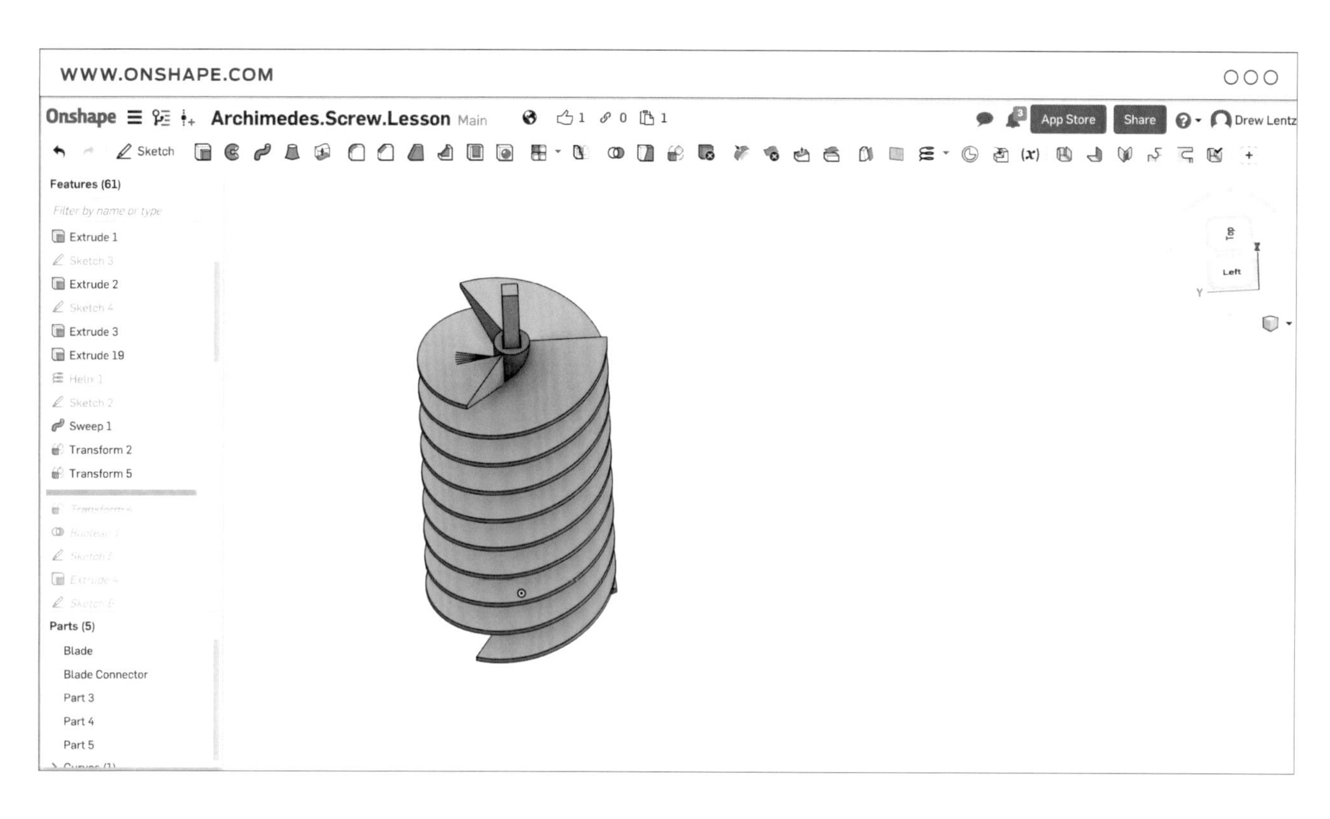

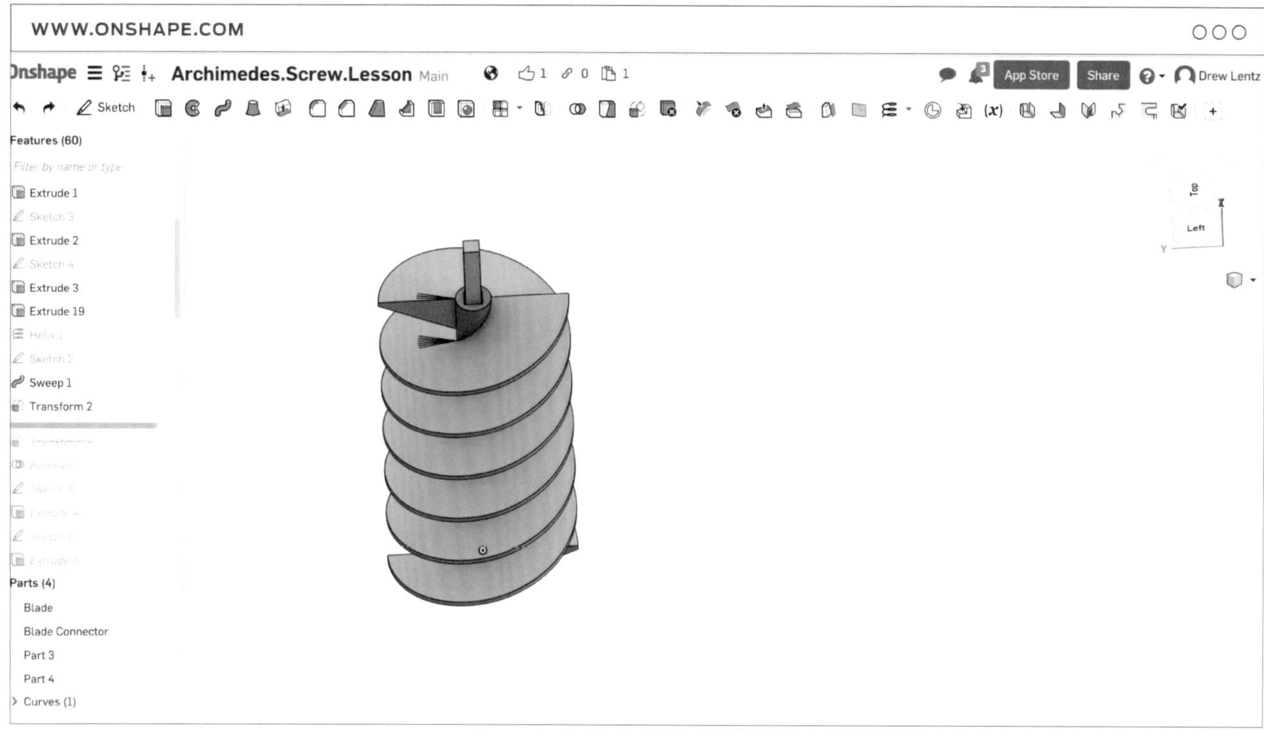

E. Edit the blade part. The blade was created using the **sweep tool** with the blade profile sketch (region to sweep) and the helix sketch (sweep path). To create a second blade using the **transform** tool, duplicate the blade and rotate 180 degrees. Experiment with creating more than 2 blades using the **transform** tool. Once you're done, merge the blade parts with the center cylinder using the **boolean** tool.

TIP: This step is only necessary if you want to create multiple blades.

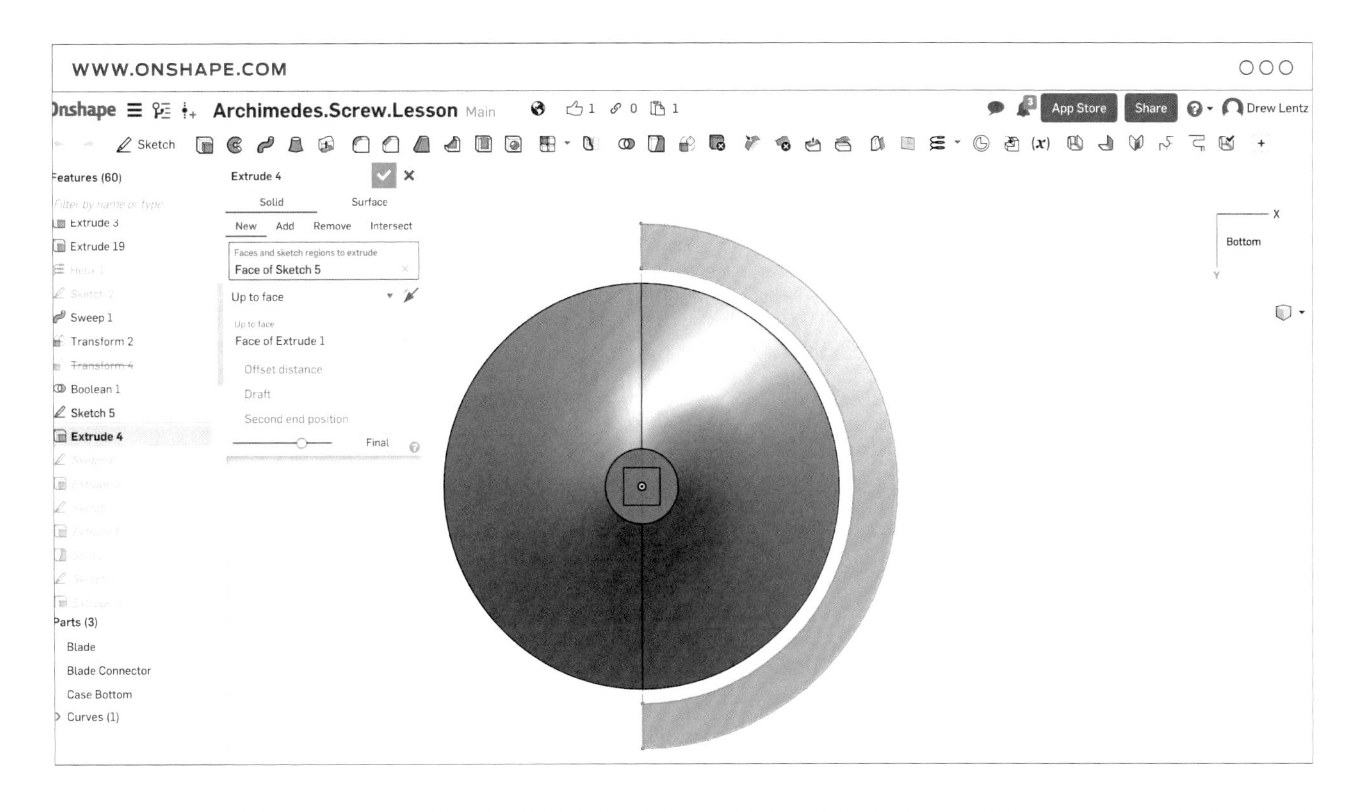

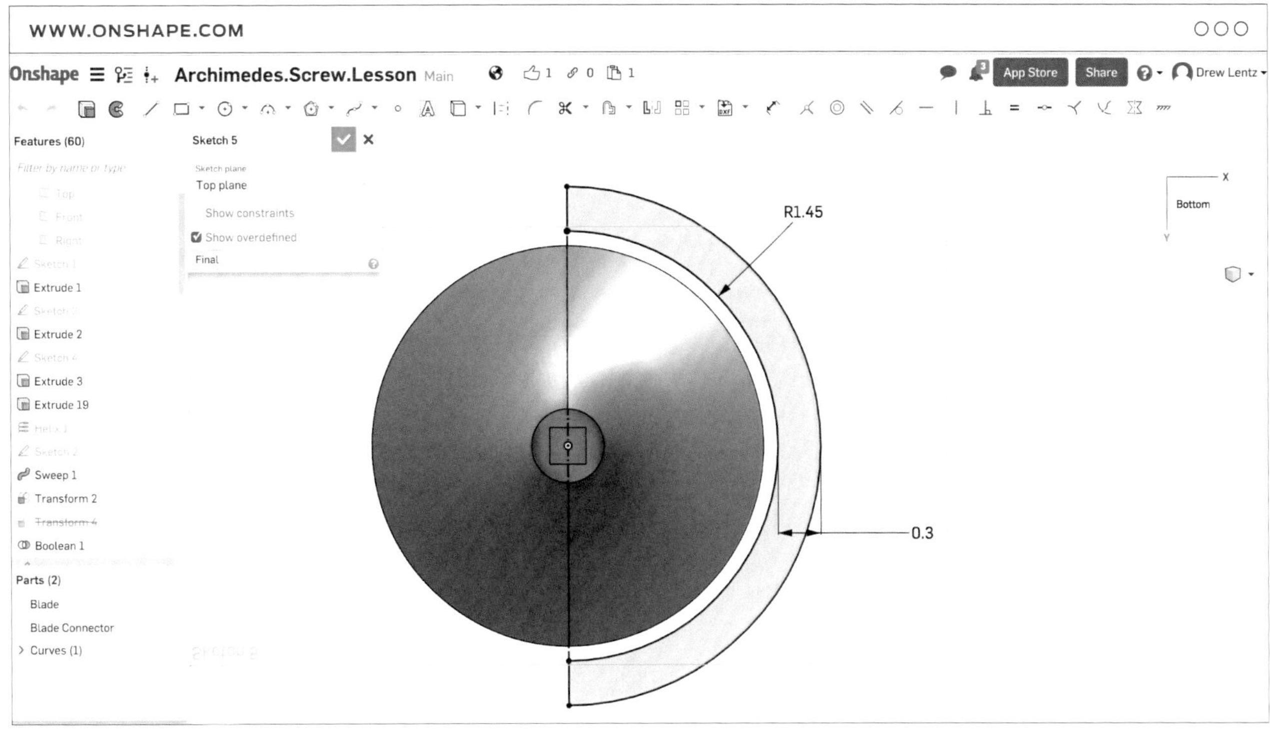

F. Edit the case sketch. This sketch defines the size and shape of the case part. Experiment with the space between the blade and the case (sample is at 0.1"). Shrinking the space will help to minimize material loss but could lead to rubbing between the blade and the case.

STEP 03:
DESIGN, PRINT, TEST, AND ITERATE

This step generally takes 3-5 class periods.

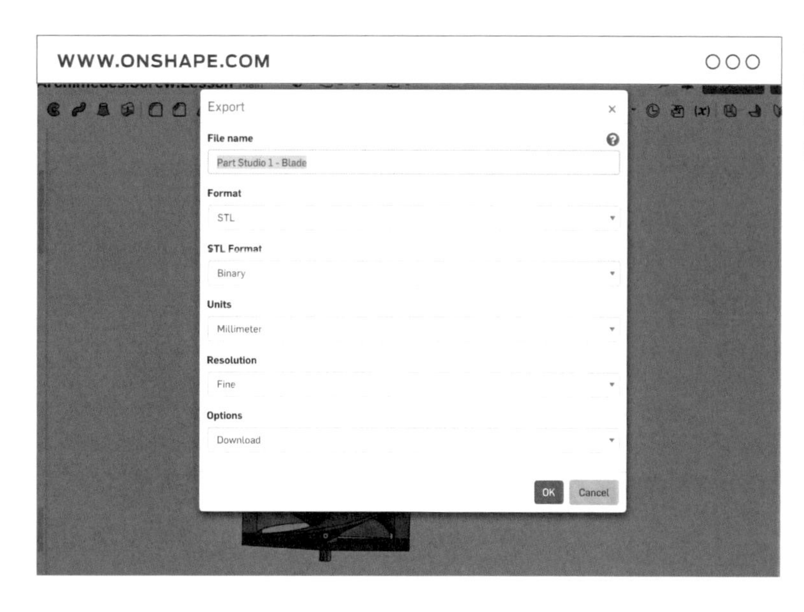

A. Export .STL files from Onshape: Right click on the part in the bottom left hand side and click **export**. Make sure to change the units to millimeters before exporting.

B. Import your files into MakerBot Print and prepare for printing.

C. Print your models:

Print Settings:

Rafts	Yes
Supports	Depends on the model
Resolution	0.2mm
Infill	5%

TIP: Given that these are large prints, test small sections or small scale versions before printing out your entire assembly.

D. Finalize your design, assemble it, and get in any last-minute tests before the official testing.

PROJECT COMPLETE:
GET CRANKIN'!

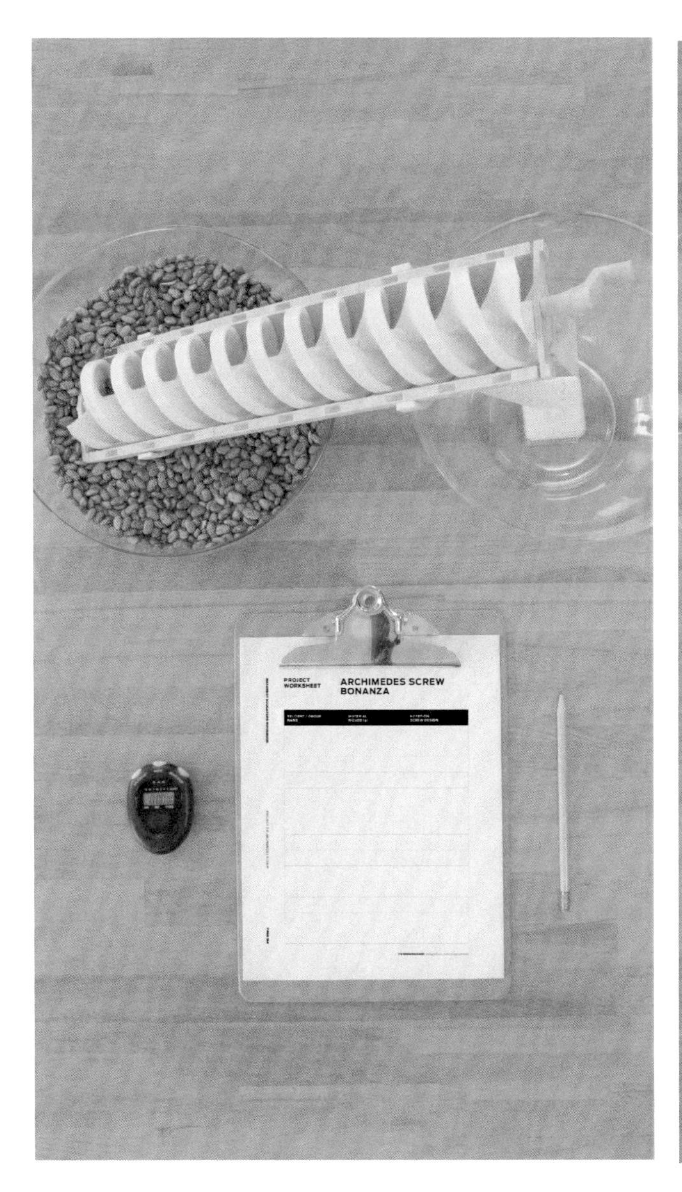

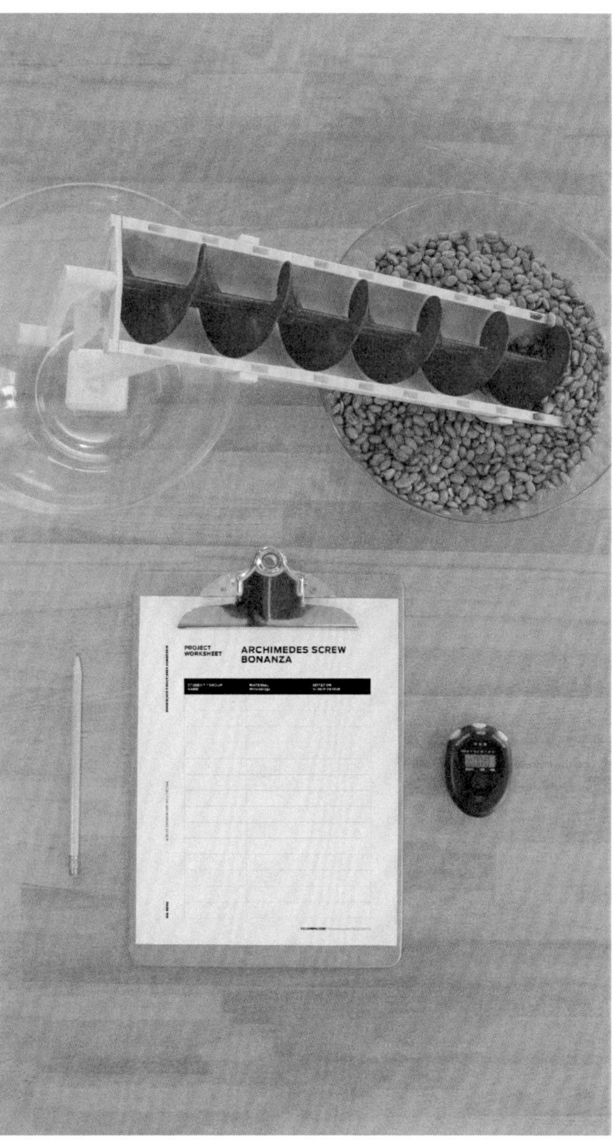

A. Each group should demonstrate their design by showing how much material they can transport from the lower bowl to the upper bowl in 30 seconds.

B. Record the results for each team, while also taking note of the differences in designs between groups.

C. Discuss the different designs from each group:

› Which worked best?

› Which adjustments proved most effective?

› If you were to redesign, what would you change?

PROJECT
WORKSHEET

ARCHIMEDES SCREW BONANZA

STUDENT / GROUP NAME	MATERIAL MOVED (g)	NOTES ON SCREW DESIGN

TO DOWNLOAD: *thingiverse.com/thing:1769714*

GOING FURTHER

A. Provide each group a different material (i.e. water, gravel, cereal, etc.) and have them think about what modifications would need to be made to accommodate these different materials.

B. If you don't want to involve 3D design in this project, there are great Customizer designs on Thingiverse that allow you to create custom screws without needing to actually design them.

PART 03:

GOING FURTHER

GOING FURTHER

POST-PROCESSING YOUR 3D PRINTS

SANDING YOUR 3D PRINTS

3D printing can bring your ideas to life, but sometimes your ideas don't stop with 3D printed parts. What if your ideas call for multiple materials? Or a smoother surface? What if your ideas are larger than the build volume of your 3D printer?

Post processing isn't necessary, but finishing techniques like sanding, gluing, painting, and silicone molding can refine your prints and take your ideas far beyond the build plate.

POST-PROCESSING TECHNIQUES:

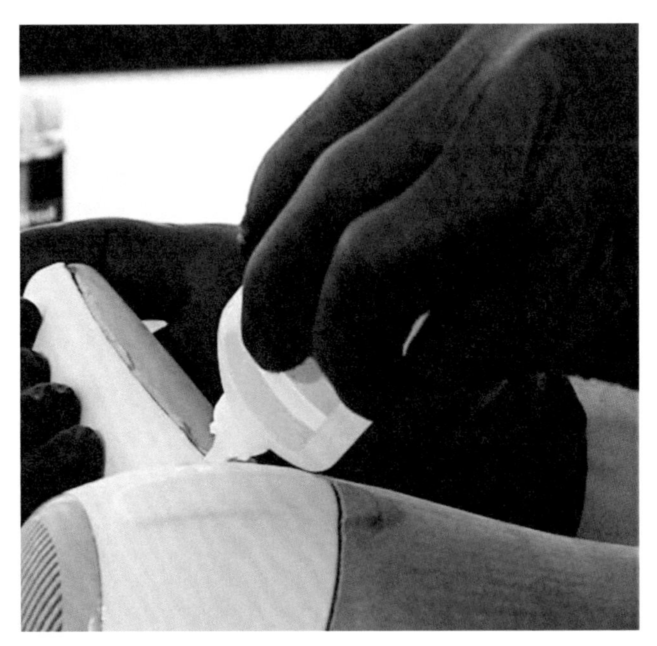

GLUING YOUR 3D PRINTS

PAINTING YOUR 3D PRINTS

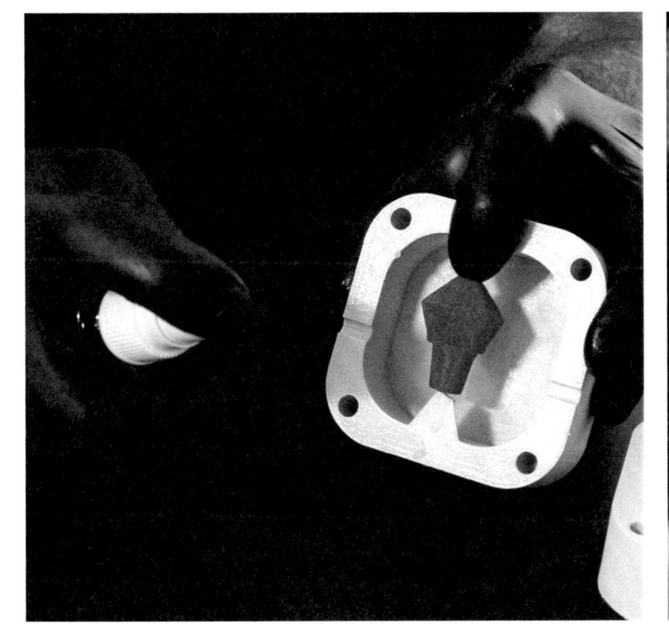

SILICONE MOLDING WITH 3D PRINTED MASTERS

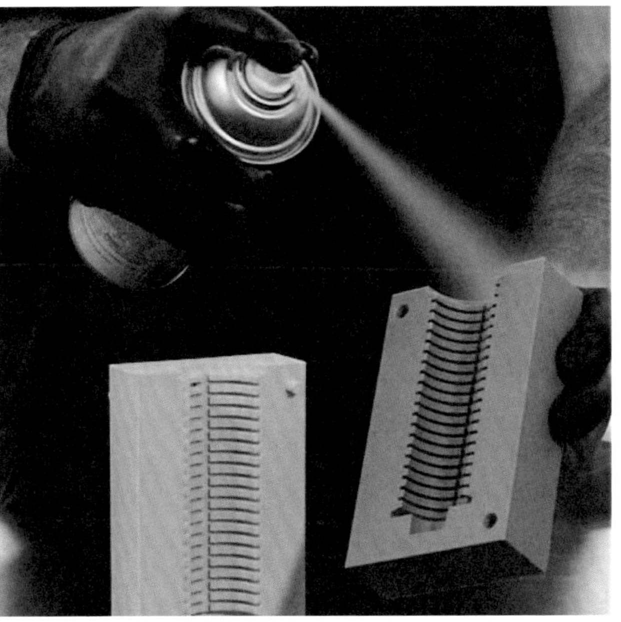

SILICONE MOLDING WITH 3D PRINTED MOLDS

POST-PROCESSING

SANDING YOUR 3D PRINTS

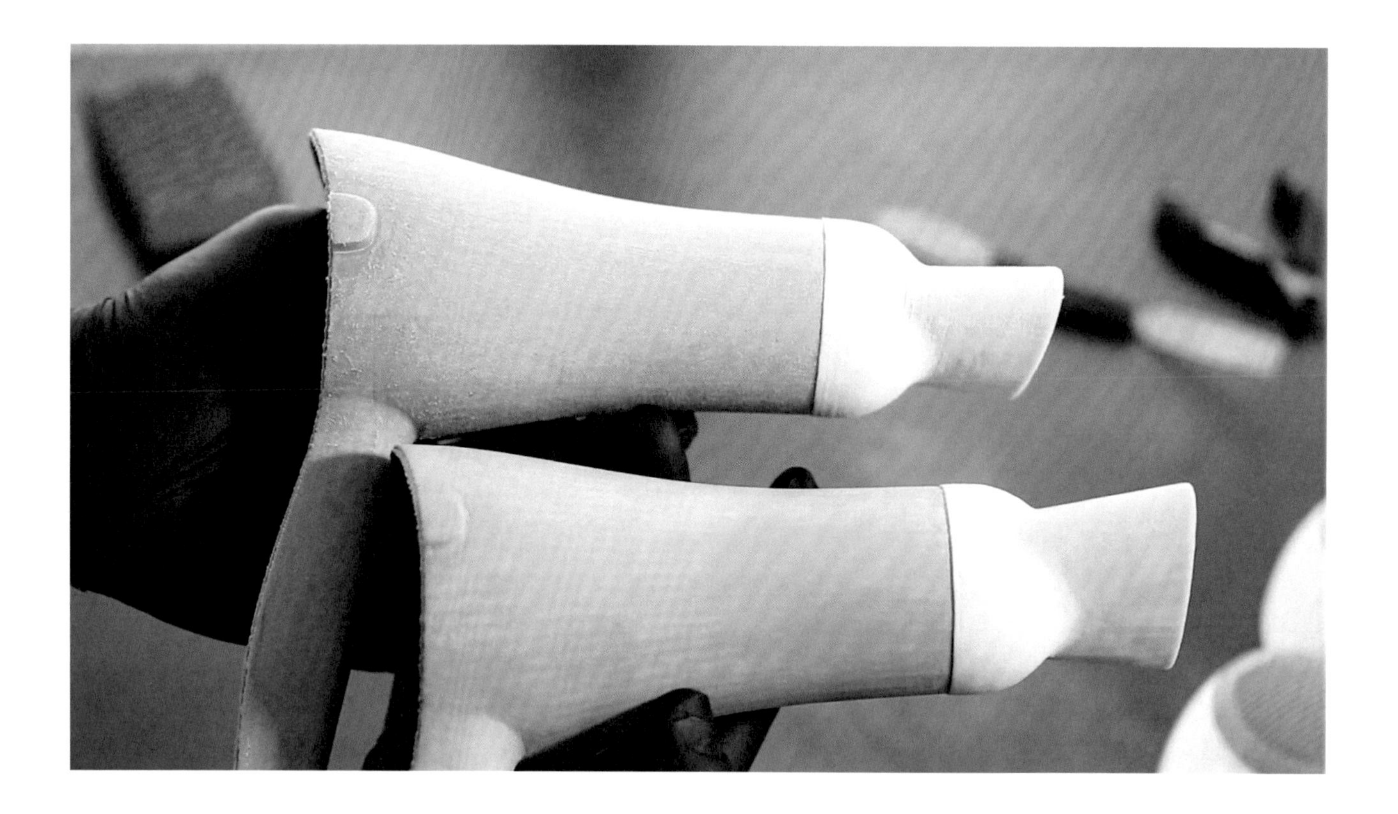

When 3D printing with FDM 3D printers, there is often a subtle layer pattern on prints. Sanding 3D prints is a great way to eliminate layer lines in preparation for painting, silicone molding, or assembly.

VIDEO TUTORIAL:
makerbot.com/professionals/post-processing/sanding

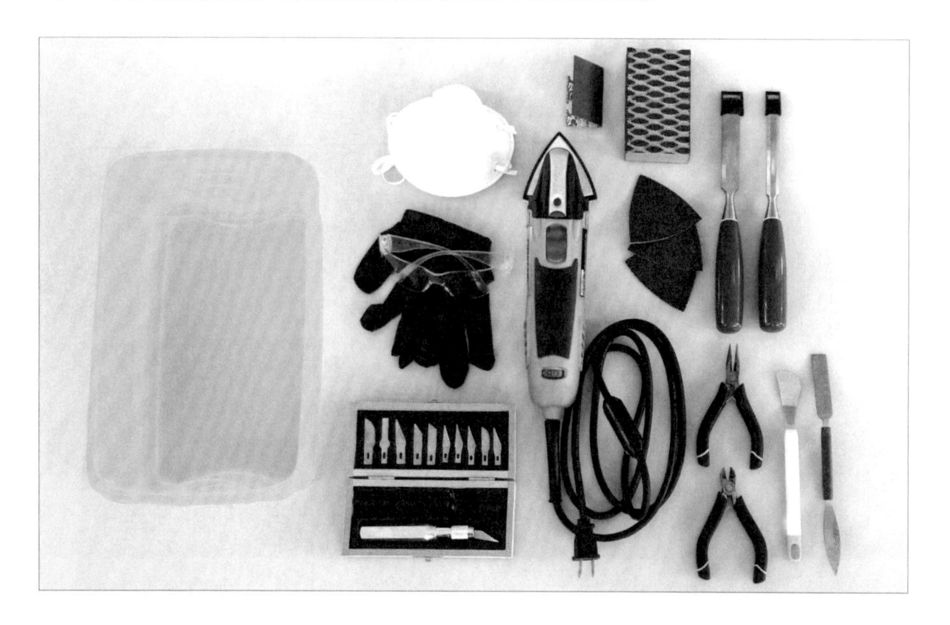

MATERIALS

Sandpaper (80, 120, 600, 1000, 1500 grit)

Handheld electric sander

Hobby knife

Needlenose pliers

Flush cutters

Chisels

Bin to hold water

Gloves, eye protection and respiratory mask

Be sure to consult the safety precautions listed by the manufacturer on any materials and equipment used in this tutorial before starting.

STEP 01: PREPARE AND PRINT YOUR MODEL

A. Consider print settings, print orientation, number of build plates.

TIP: Surfaces printed vertically will have the smoothest finish.

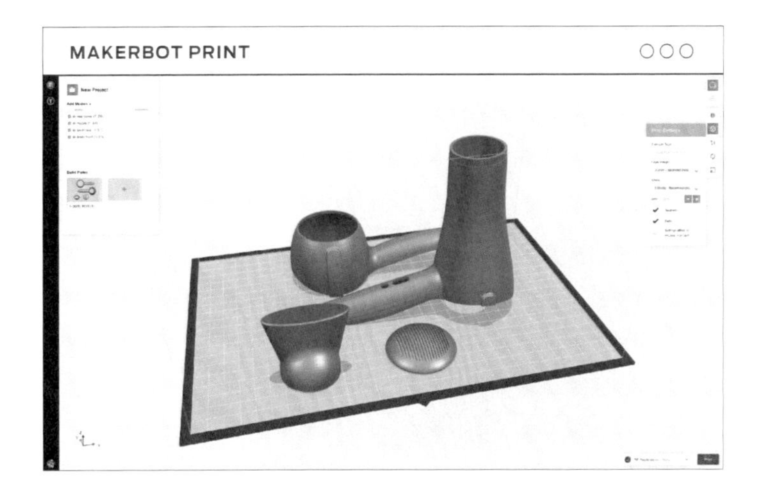

HAIR DRYER SHELL: *grabcad.com/library/hair-dryer-shell-1*

STEP 02: REMOVE SUPPORT MATERIALS AND RAFTS

A. Remove the prints from the build plate.

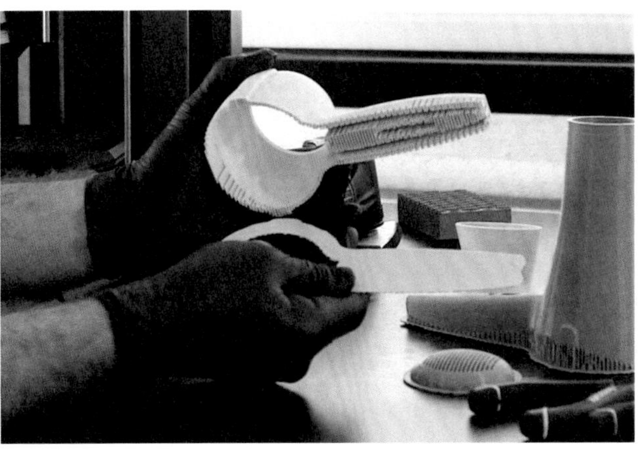

B. Remove rafts and support material from your print. Be sure to wear eye protection when removing supports using pliers or clippers.

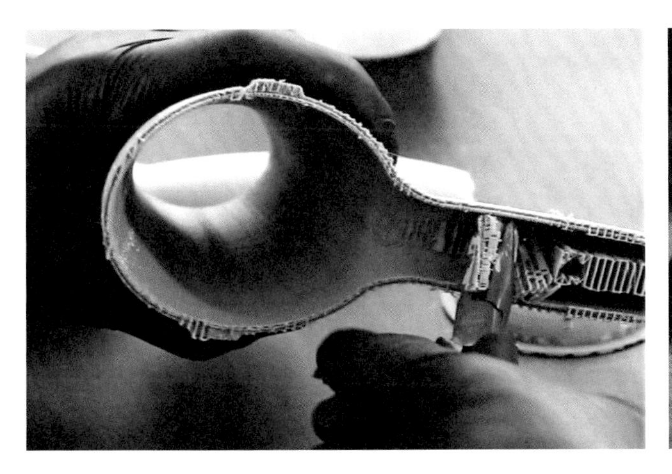

C. Remove large pieces of support first and then approach smaller pieces and fine details.

D. Clean the edges and seams of your print to ensure better alignment of pieces.

STEP 03: SAND WITH COARSE GRIT SANDPAPER (DRY)

When sanding your prints, begin with lower grit sandpaper, and move on to higher grit as the print surface gets smoother.

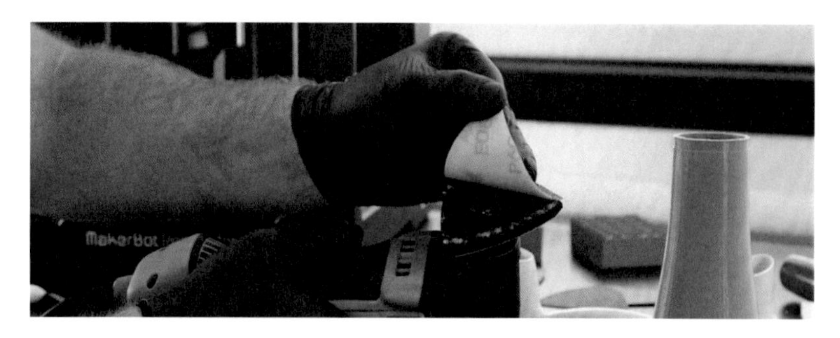

A. Begin sanding, with coarse 80 grit sandpaper.

B. When sanding with 80 grit the goal is to remove any leftover blemishes from raft or support material, and create an even surface to refine. This process will remove the most amount of material and take the most time.

C. In early stages of sanding you will notice your print's surface changing from shiny to dull. The shine will return as you move to higher grit sandpaper.

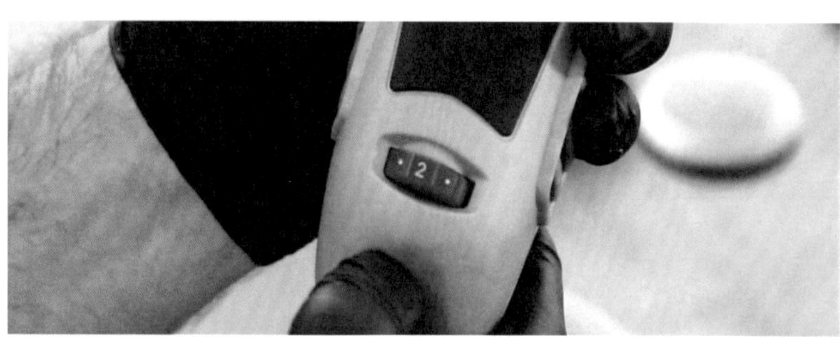

D. To save time you can use an electric sander or multitool. If using a multitool, be sure to choose a low setting to avoid overheating/deforming the PLA.

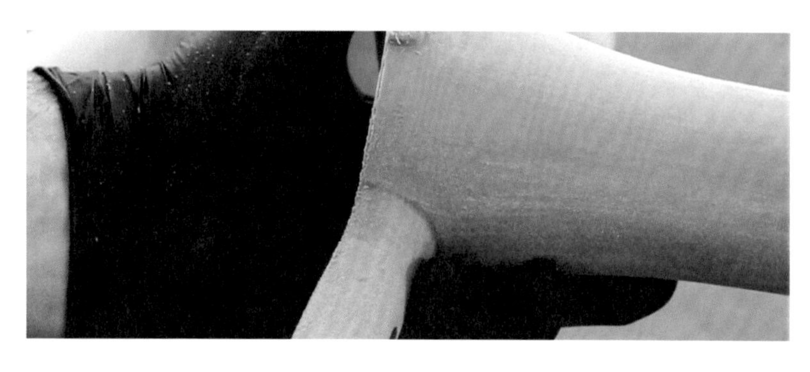

E. After each sanding stage, clean the print of any dust and inspect for a uniform surface finish. Move on to sanding with higher grit when all large-to-medium sized imperfections or blemishes have been removed.

STEP 04: SAND WITH FINE SANDPAPER (WET)

When your print's surface is even and refined it's time to wet sand the model using fine grit sandpaper. This process doesn't remove much material, but will polish its surface.

A. Submerge the print in a tub of water.

B. Sand the print using 1000 grit sandpaper until it is completely smooth to the touch.

C. Dry the print and inspect for a uniform surface finish.

TIP: Be sure not to sand in one place for too long as heat generated from friction could deform the PLA.

D. Once finished, your print's surface should now be even, blemish free, and glass-like to the touch. Here you can see the difference between the unfinished surface and final sanded part.

GLUING YOUR 3D PRINTS

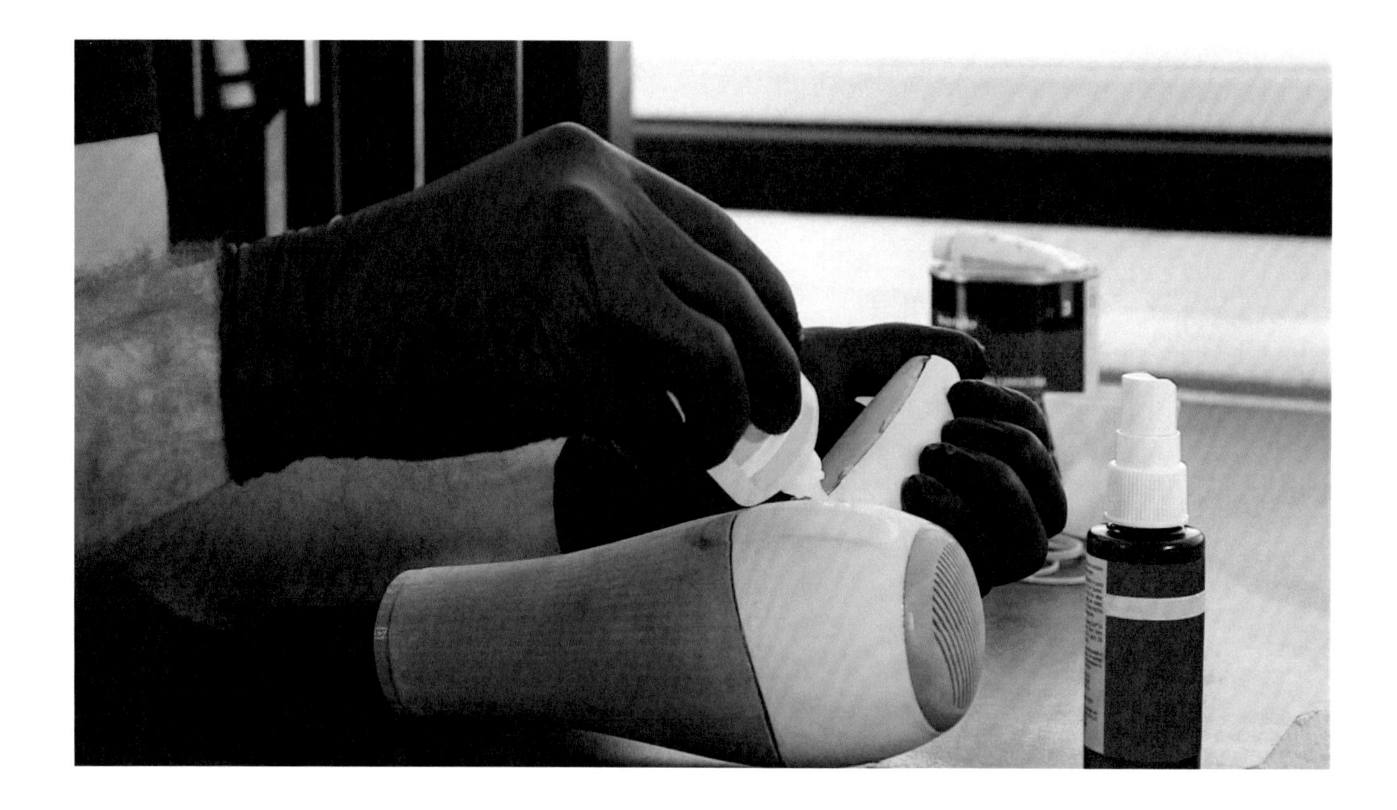

Gluing your 3D prints is essential if you're looking to combine multiple components of an assembly or to create a model larger than the build volume of your 3D printer. Working time will vary depending on your model.

VIDEO TUTORIAL:
makerbot.com/professionals/post-processing/gluing

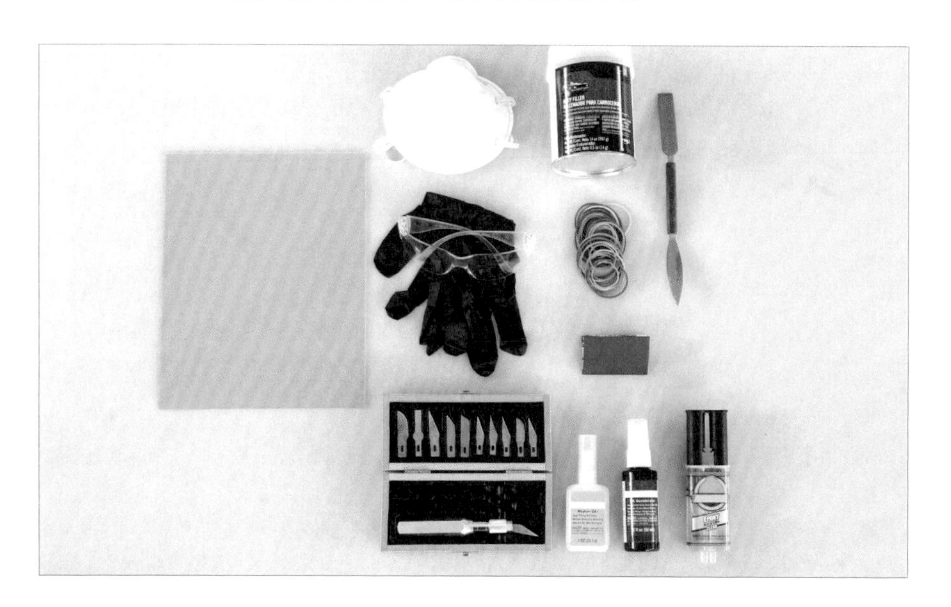

MATERIALS

- 3D print with multiple components
- Cyanoacrylate glue
- 2 part epoxy
- Rubber bands
- Paper towels
- Scraping tool
- 400-1000 grit sandpaper
- Gloves, eye protection and respiratory mask

Be sure to consult the safety precautions listed by the manufacturer on any materials and equipment used in this tutorial before starting.

IF YOU'VE ALREADY COMPLETED SANDING YOUR PRINTS, JUMP TO STEP 3

STEP 01: PREPARE AND PRINT YOUR MODEL

A. Consider print settings, print orientation, number of build plates.

TIP: Surfaces printed vertically will have the smoothest finish.

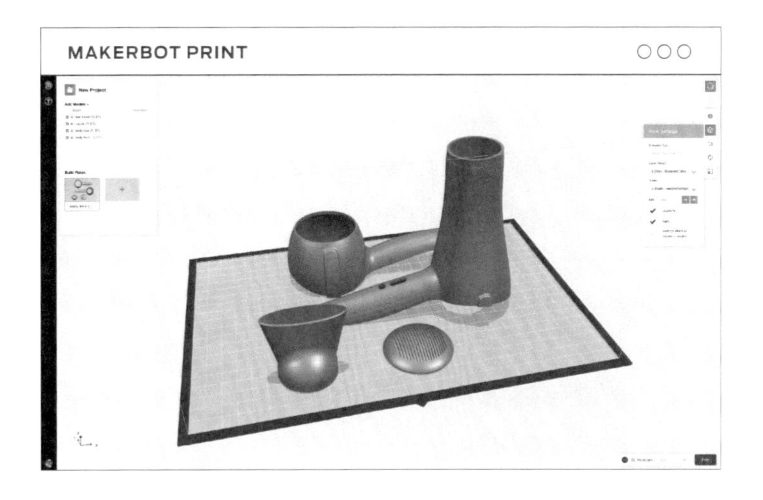

HAIR DRYER SHELL: *grabcad.com/library/hair-dryer-shell-1*

STEP 02: REMOVE SUPPORT MATERIALS AND RAFTS

A. Remove the prints from the build plate.

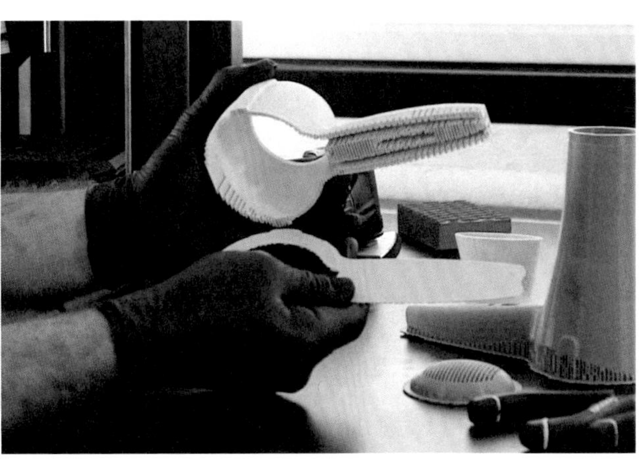

B. Remove rafts and support material from your print. Be sure to wear eye protection when removing supports using pliers or clippers.

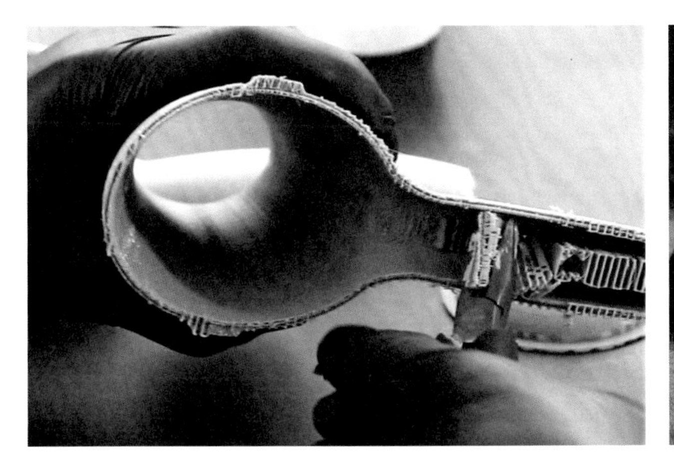

C. Remove large pieces of support first and then approach smaller pieces and fine details.

D. Clean the edges and seams of your print to ensure better alignment of pieces.

STEP 03: INSPECT JOINTS

Make sure the areas where the two pieces will be bonded are clean and clear of any debris or blemishes. Follow the steps below to ensure your print is ready for gluing.

A. Inspect your part. It's ready to glue when all surfaces and joints are smooth and fit together.

STEP 04: SECURE COMPONENTS

A. Join components together using rubber bands.

TIP: If rubber bands don't work for the geometry of your parts, try straps, tape or clamps.

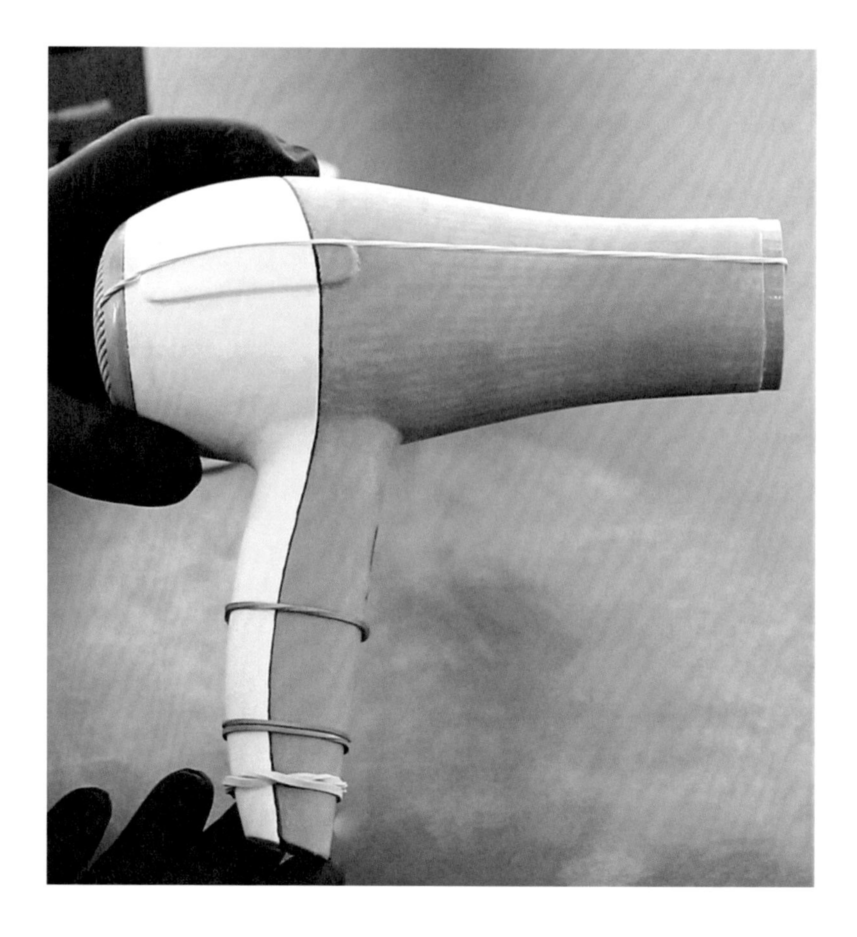

STEP 05: SPOT GLUE

Once your parts' surfaces have been prepared and are in a fixed and stable position, you can begin gluing.

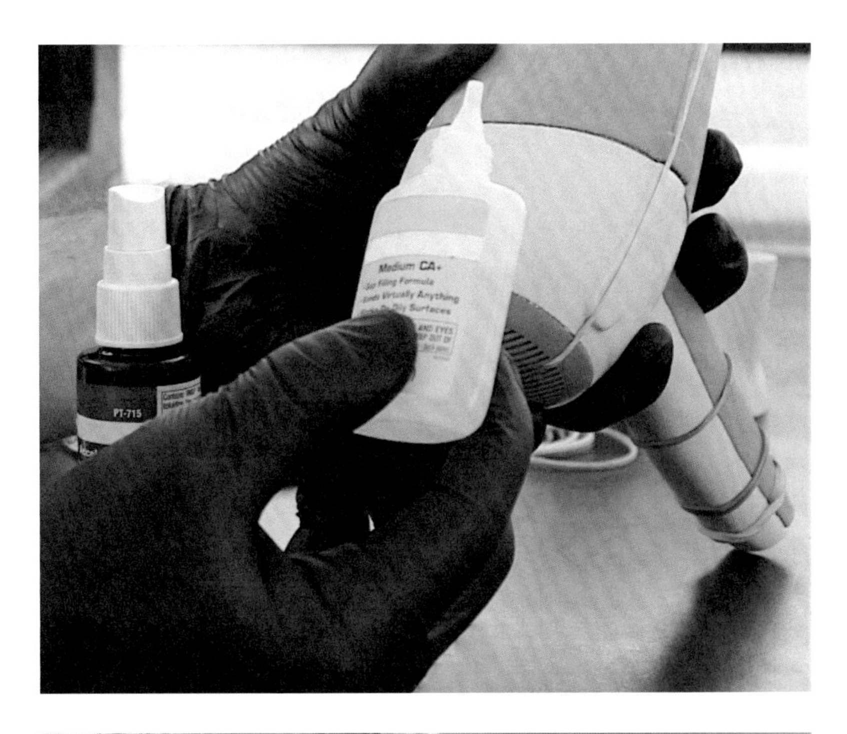

A. Begin gluing by applying small amounts of glue to central locations around your model. This will ensure that your parts are evenly attached in all places.

TIP: Work from the center outwards in regular intervals for even gluing.

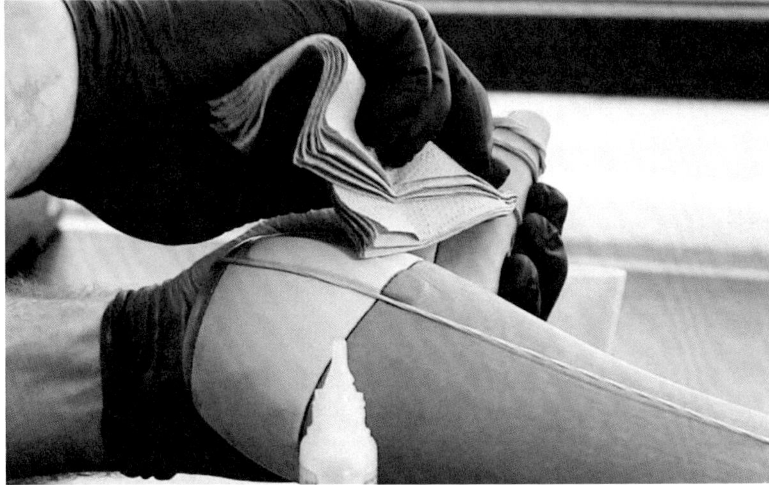

B. Wipe away any excess glue with a paper towel.

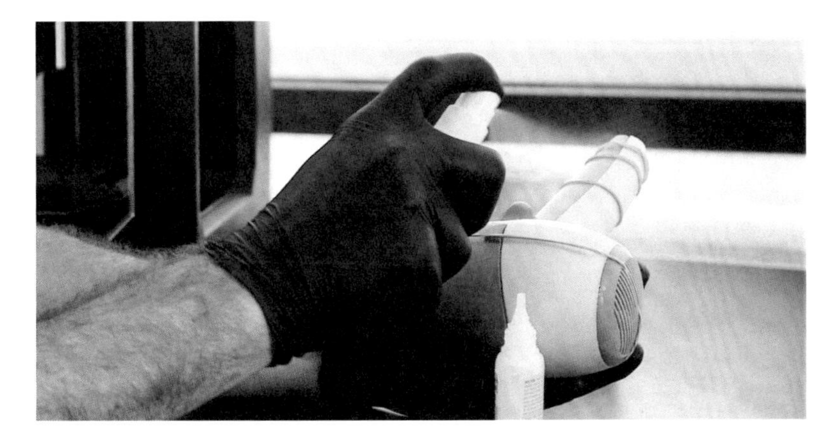

C. Spray accelerant to speed up the drying process (optional).

STEP 06: GLUE ALL SEAMS

Once the seams are correctly aligned, apply glue to each seam around your assembled print.

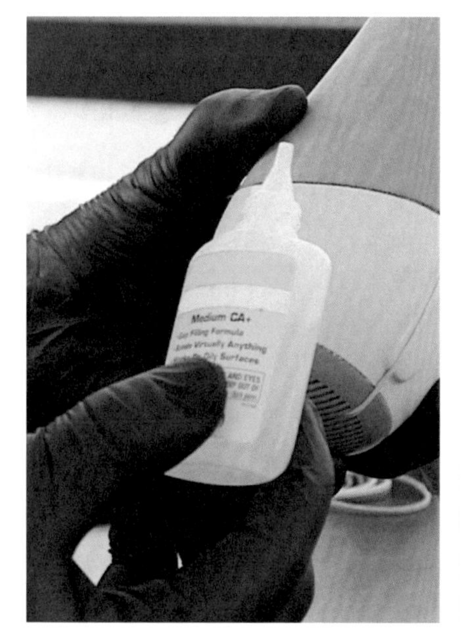

A. Apply glue evenly to the remaining unglued seams.

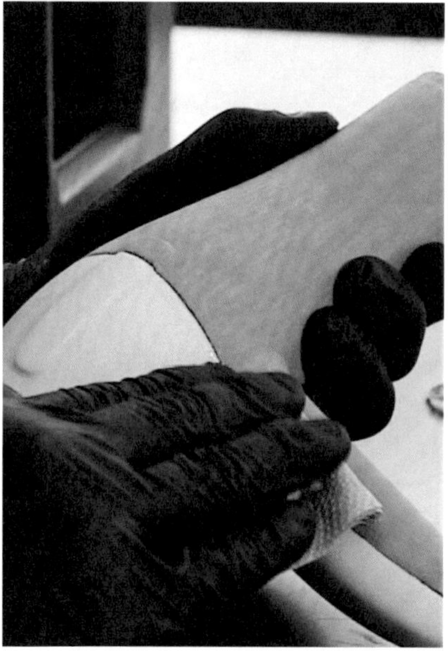

B. Wipe away any excess glue.

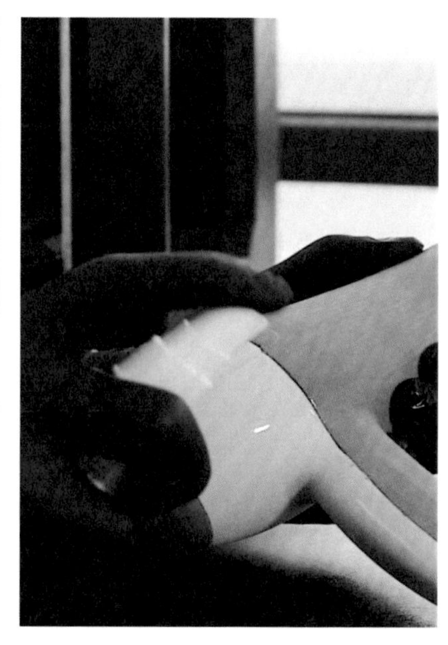

C. Spray accelerator to speed up the drying process (optional).

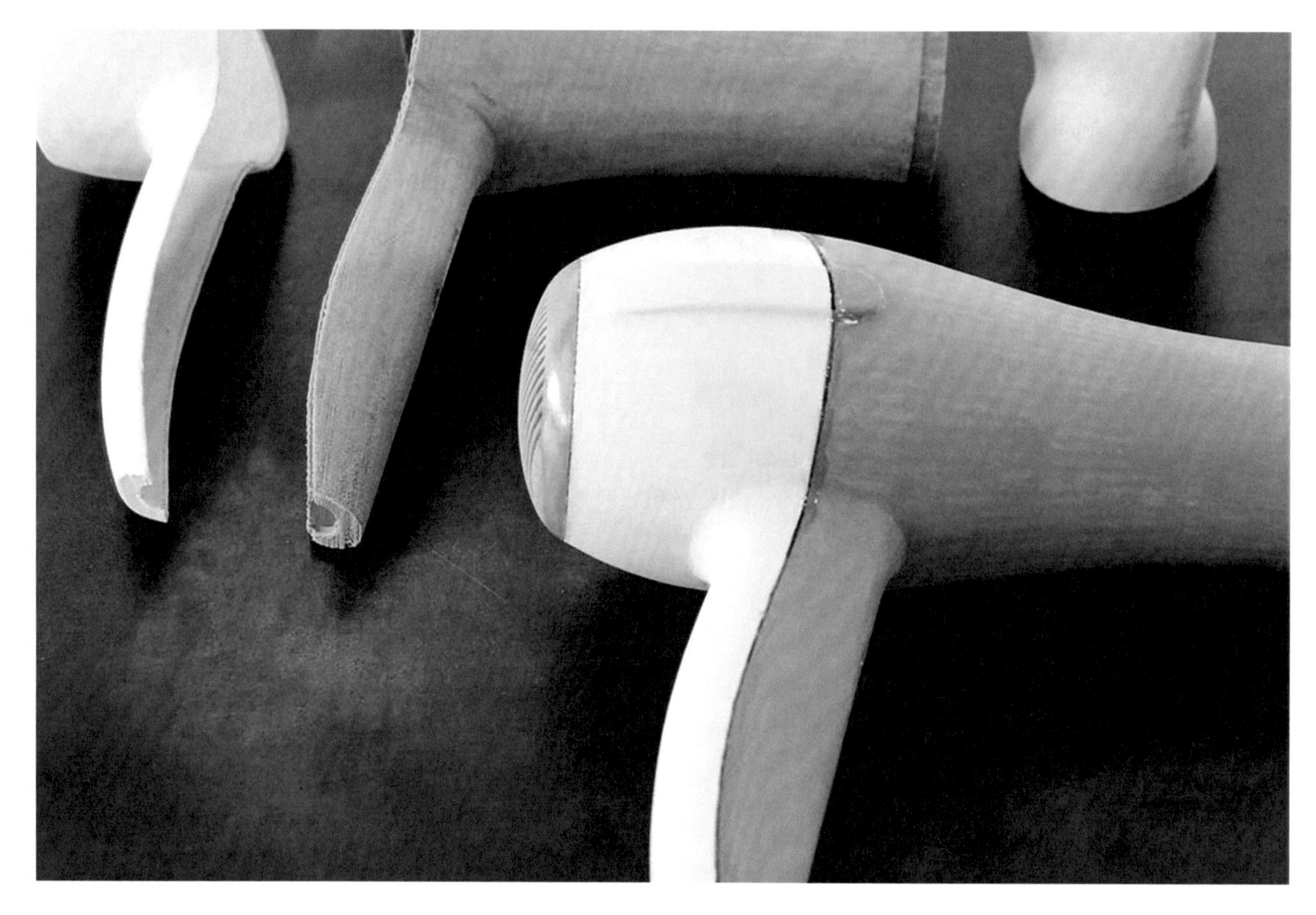

D. Inspect the part to make sure each seam is fully glued.

STEP 07: (OPTIONAL) FILL SEAMS

If you find the seams are rough or have gaps, you can use a body filler to smooth them - this works especially well if you plan to paint your model or create a silicone mold from it. Body fillers come in two parts - filler and hardener.

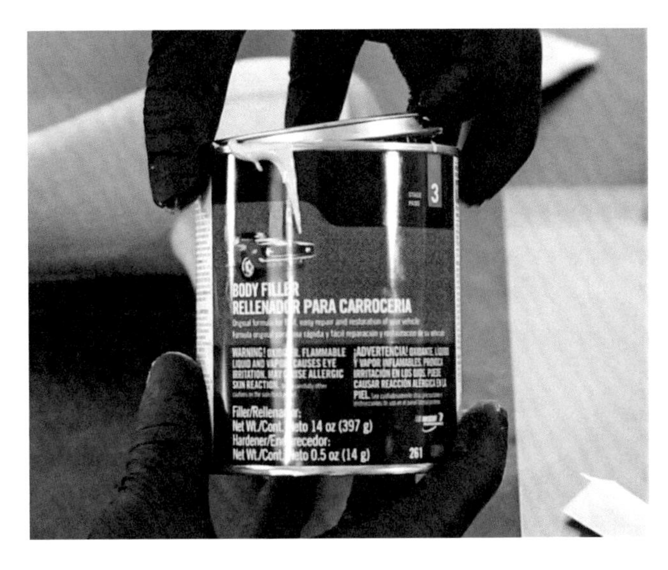

A. Pour a small amount of filler on a tray.

B. Spread a small amount of hardener next to filler.

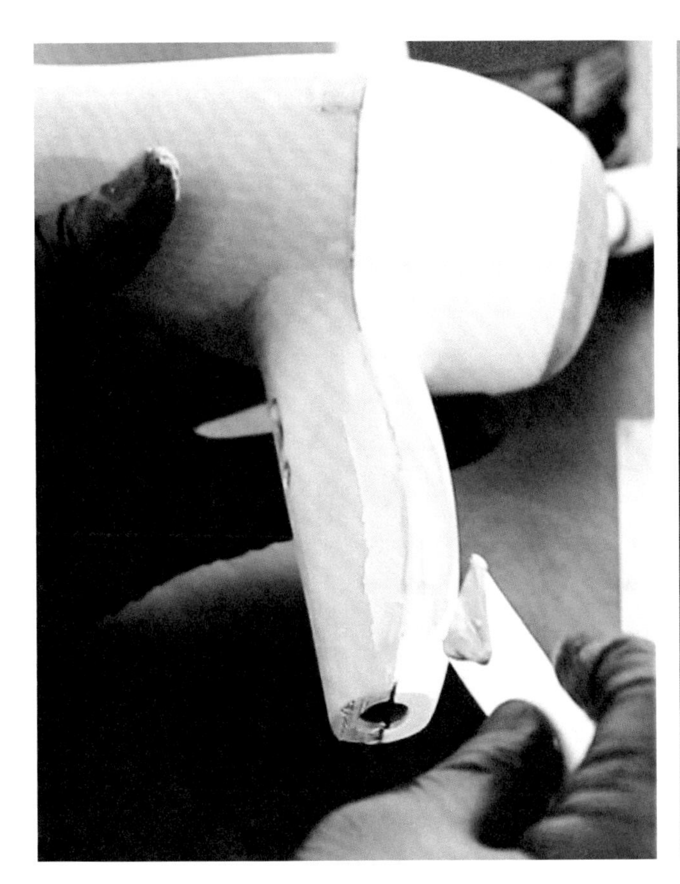

C. Mix and generously spread across seams of the assembled print.

D. Allow time to cure according to your filler's instructions.

POST-PROCESSING

PAINTING YOUR 3D PRINTS

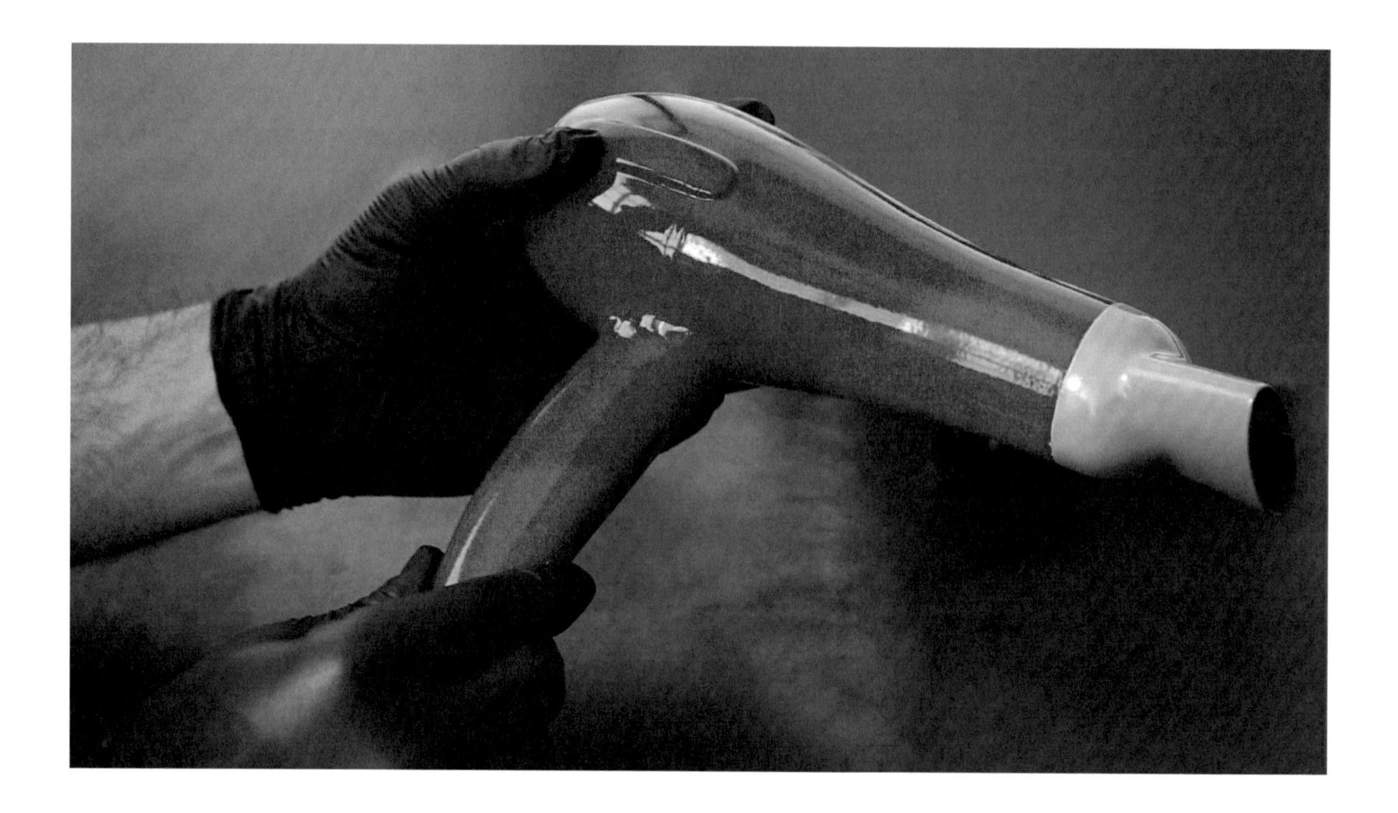

If you're looking to add a bit more creativity into your classroom 3D printing activities, painting your models is a great solution.

VIDEO TUTORIAL:
makerbot.com/professionals/post-processing/painting

MATERIALS

- 3D print
- Rubbing alcohol
- Paper towels
- Spray primer/filler
- Hanging cord
- 80, 120, 240, 600, 1000 grit sandpaper
- Needlenose pliers
- Sanding block and sandpaper
- Gloves, eye protection and respiratory mask

Be sure to consult the safety precautions listed by the manufacturer on any materials and equipment used in this tutorial before starting.

IF YOU'VE ALREADY COMPLETED SANDING AND GLUING YOUR PRINTS, JUMP TO STEP 4

STEP 01: PREPARE AND PRINT YOUR MODEL

A. Consider print settings, print orientation, number of build plates.

TIP: Surfaces printed vertically will have the smoothest finish.

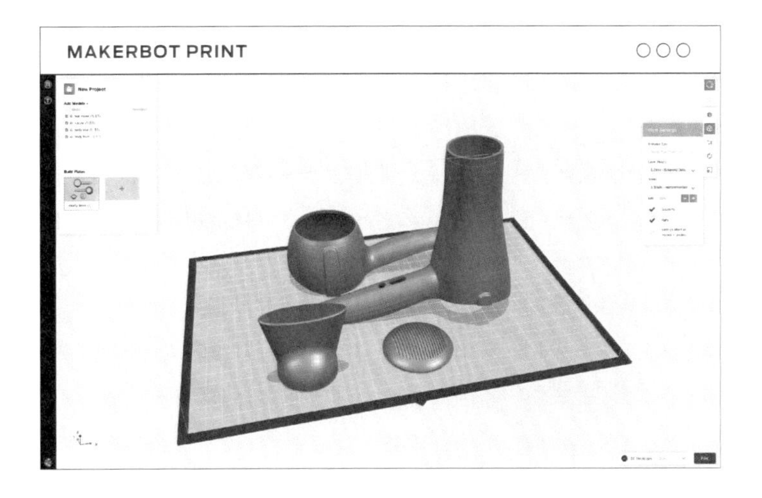

HAIR DRYER SHELL: *grabcad.com/library/hair-dryer-shell-1*

STEP 02: REMOVE SUPPORT MATERIALS AND RAFTS

A. Remove the prints from the build plate.

B. Remove rafts and support material from your print. Be sure to wear eye protection when removing supports using pliers or clippers.

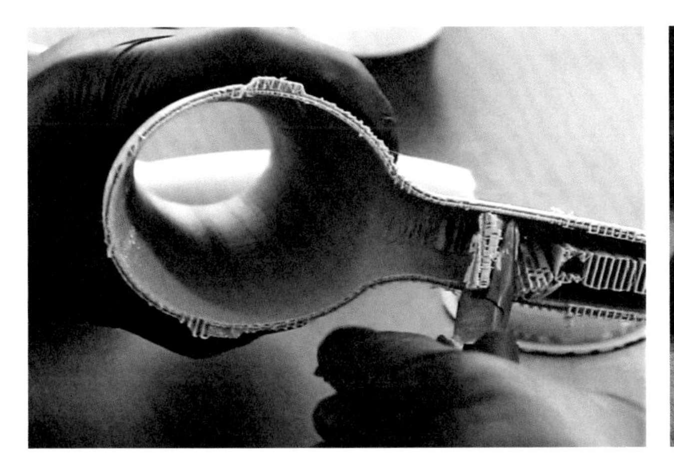

C. Remove large pieces of support first and then approach smaller pieces and fine details.

D. Clean the edges and seams of your print to ensure better alignment of pieces.

STEP 03: SAND, GLUE AND USE FILLER (OPTIONAL)

For more detail on gluing and sanding, see previous sections.

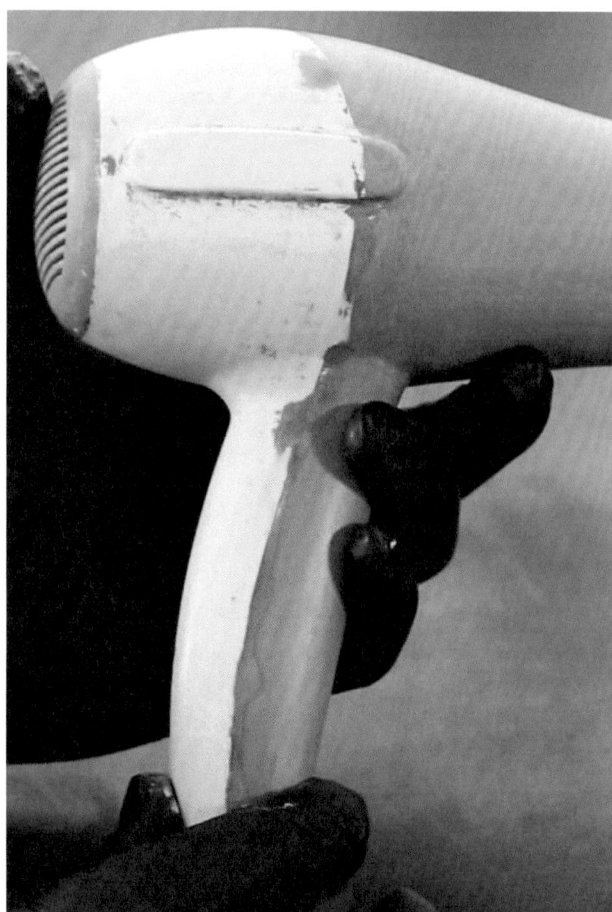

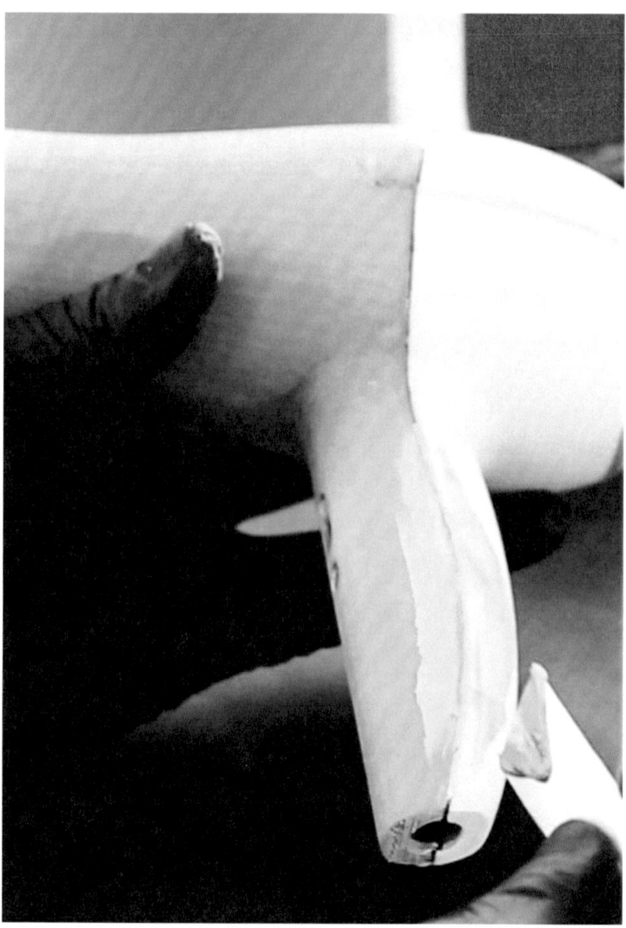

STEP 04: HANG YOUR PRINTS

Before painting, hang your prints in an open, dust-free space with plenty of ventilation. This will allow you to paint all surfaces evenly without having to handle the part while paint is drying.

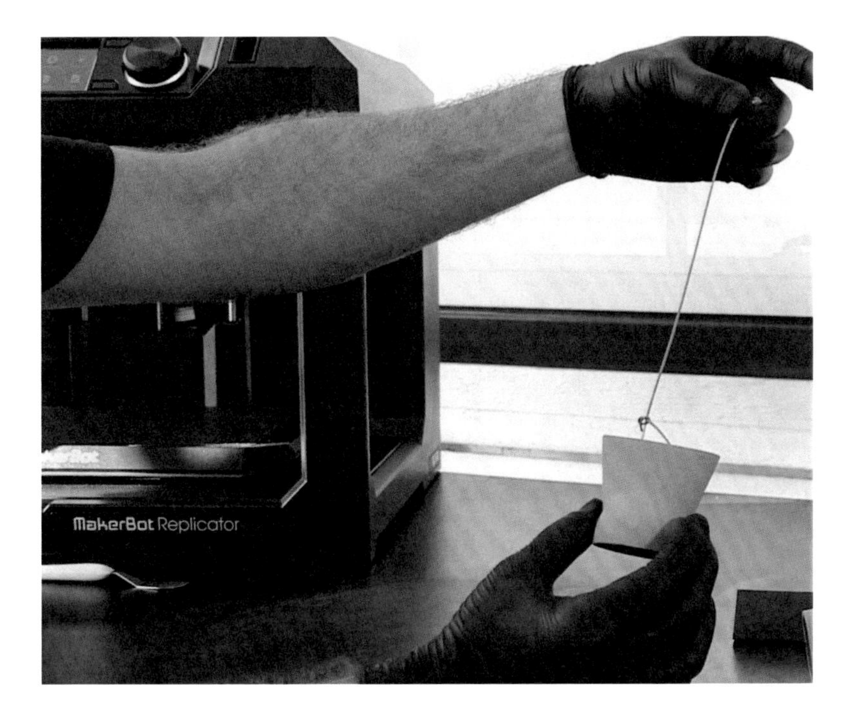

A. Tie hanging cord around the print.

B. Hang print in an open room.

STEP 05: APPLY PRIMER/FILLER

A few layers of primer/filler will fill any small surface imperfections before painting.

A. Apply primer filler using wide strokes beginning and ending in the space outside of the part. Paint in thin layers.

B. Once you've sprayed 2-3 layers of primer/filler, allow 30-40 minutes for drying.

STEP 06: SAND

A. Lightly sand with 1000 grit sandpaper (dry). This will smooth the surface of the primer filler.

B. Look for surface imperfections you would like to smooth. Continue to apply layers of primer/filler, if needed.

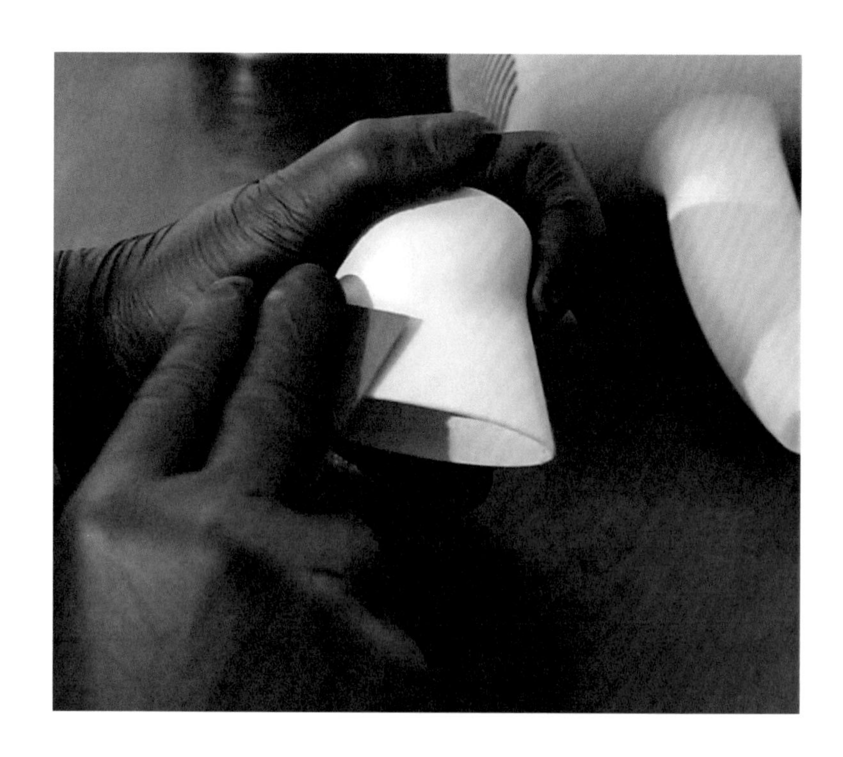

STEP 07: PAINT

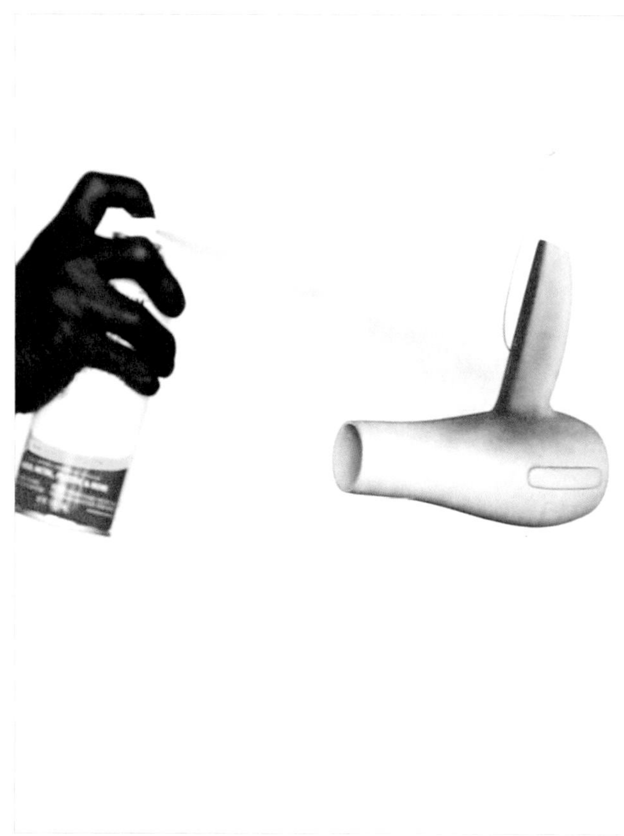

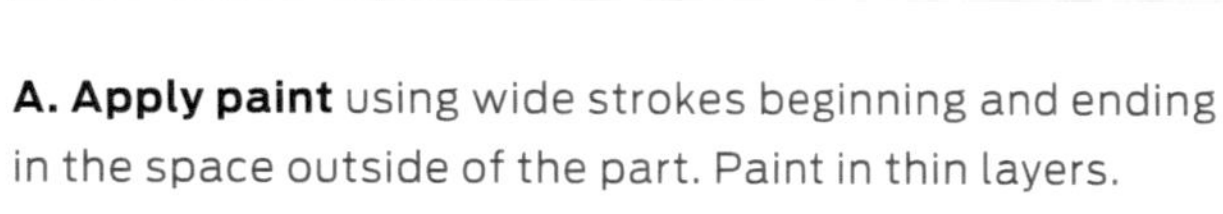

A. Apply paint using wide strokes beginning and ending in the space outside of the part. Paint in thin layers.

B. Let paint cure overnight.

TIP: When choosing paint, make sure to pick something that adheres well to plastic.

STEP 08: THE FINAL PRODUCT

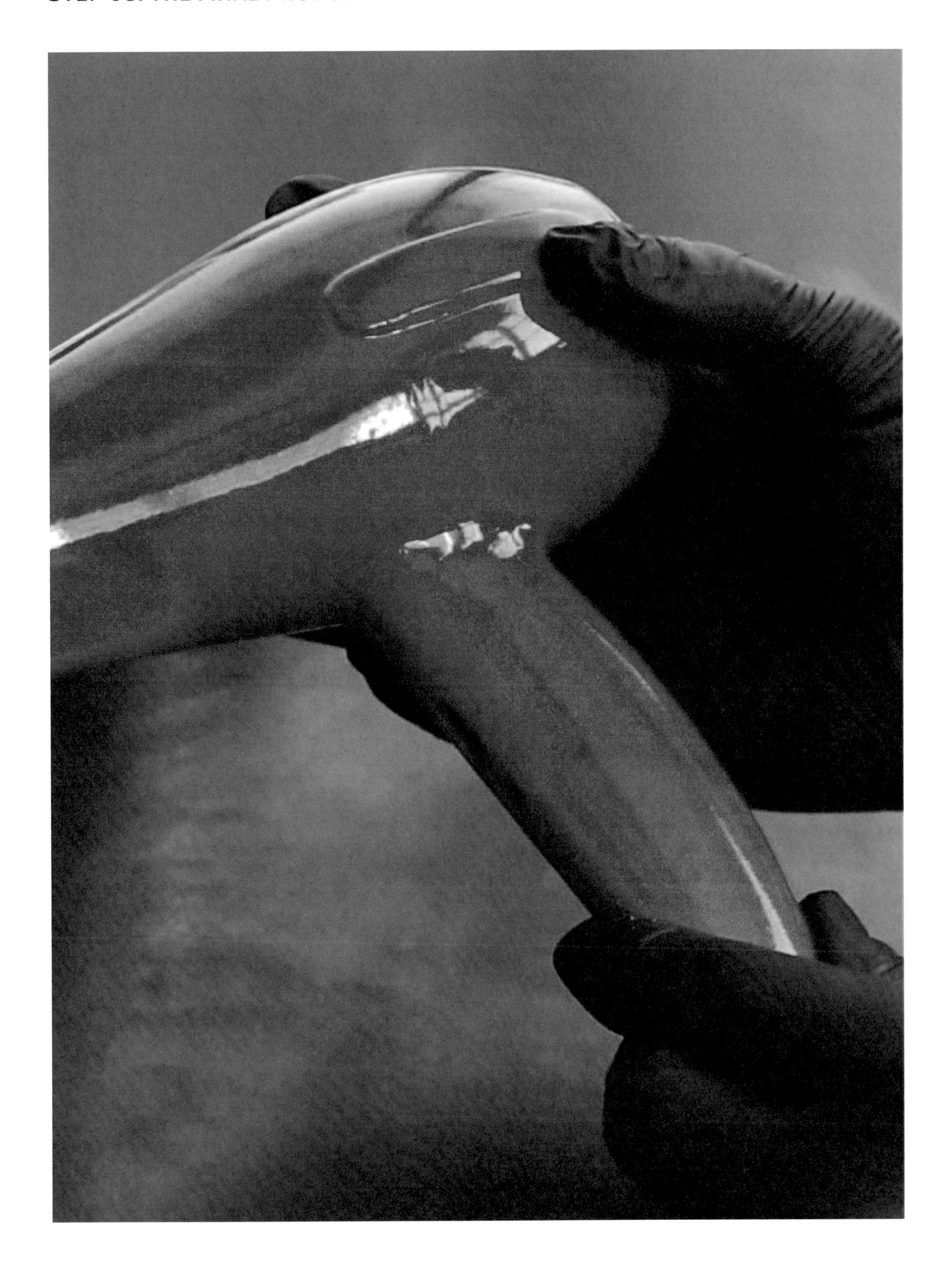

POST-PROCESSING

SILICONE MOLDING WITH 3D PRINTED MASTERS

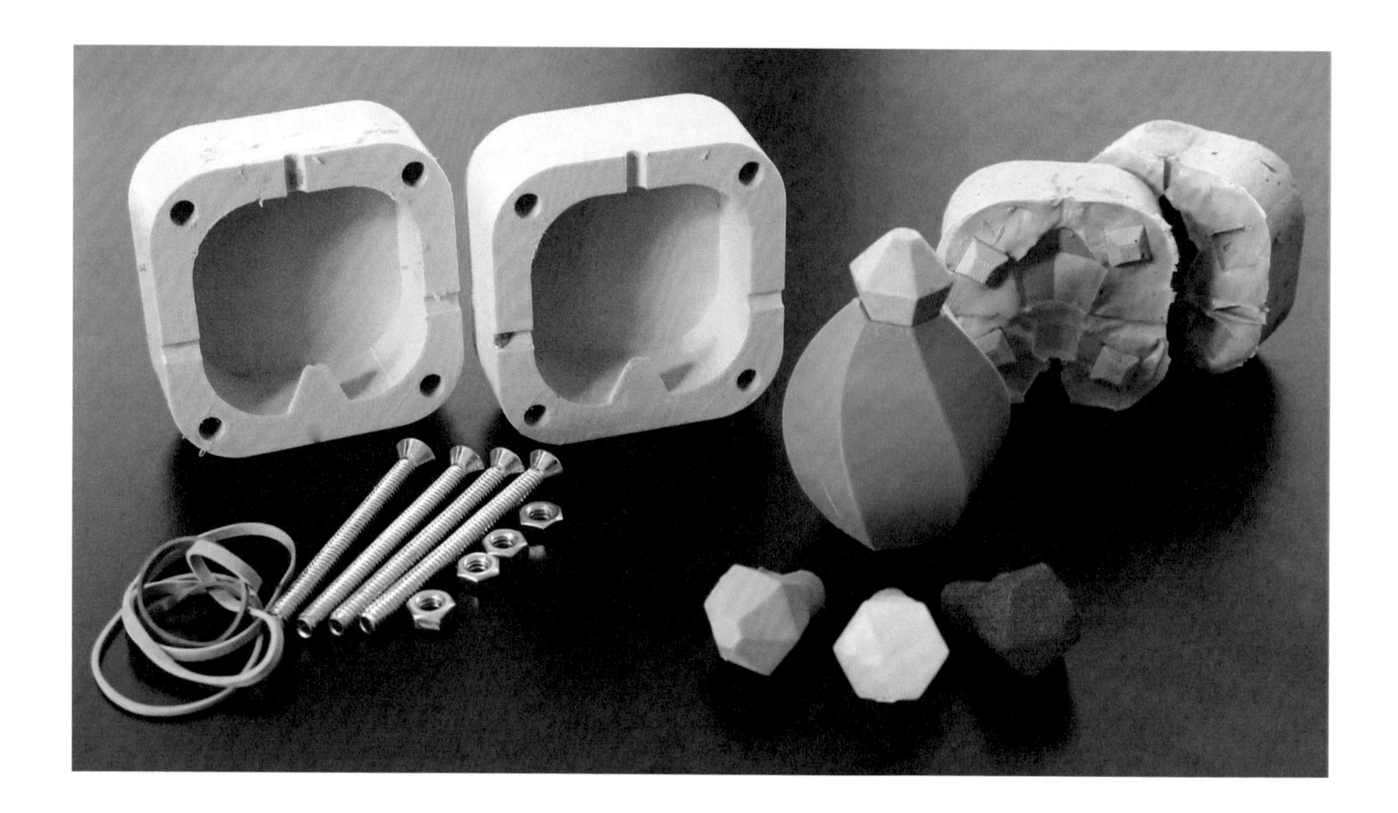

You don't have to be an engineer, designer, or know about molding and casting to make a silicone mold. Silicone molding is an inexpensive, easy, fun way to create objects in a number of different materials, or create a mold around a 3D printed part.

VIDEO TUTORIAL:
makerbot.com/professionals/post-processing/silicone-molding-1

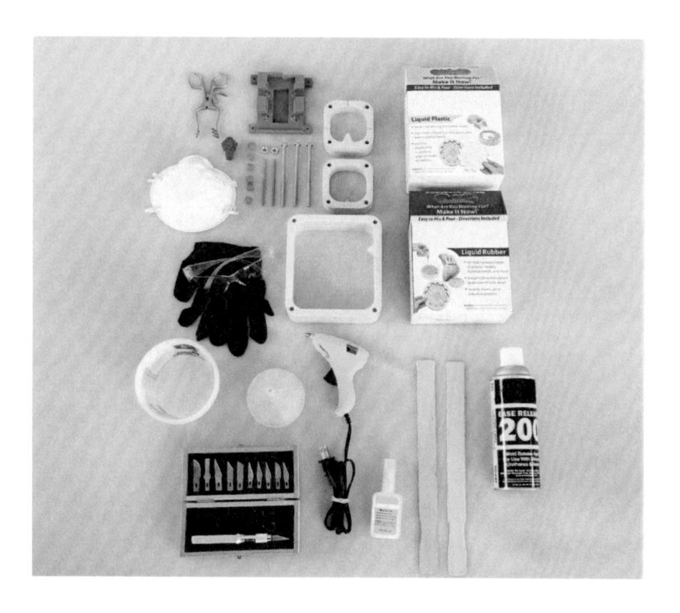

MATERIALS

- 3D printed mold box, hardware, vents & keys
- Master (the print you are molding)
- Silicone
- Resin and dye
- Mixing cups
- Mixing sticks
- Mold release spray
- Hot glue or cyanoacrylate glue
- Funnel
- Hobby knife
- Rubber bands, tape, or straps
- Gloves, eye protection and respiratory mask

Be sure to consult the safety precautions listed by the manufacturer on any materials and equipment used in this tutorial before starting.

STEP 01: CHOOSE A FILE TO CAST A MODEL AROUND

A. Print a part that you would like to either make several of, or create in a material not supported by your printer. Here, we chose the cap to a perfume bottle.

STEP 02: CREATE YOUR MOLD BOX

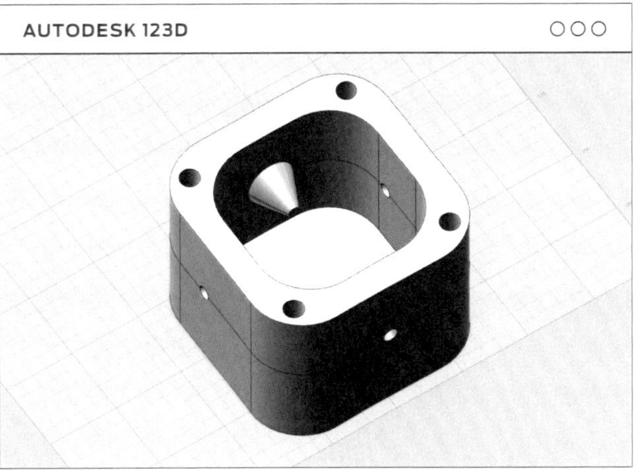

A. Design a mold box to hold the silicone in place around your part when pouring. Your master will need to be suspended in this structure. You can create mold boxes from foam core board, legos, or 3D printing.

Designing and 3D printing mold boxes allows you to create custom pour holes and vents, easily calculate the volume of the mold, create boxes that perfectly fit the parts you plan to create a mold of, and re-use mold boxes to create multiple molds.

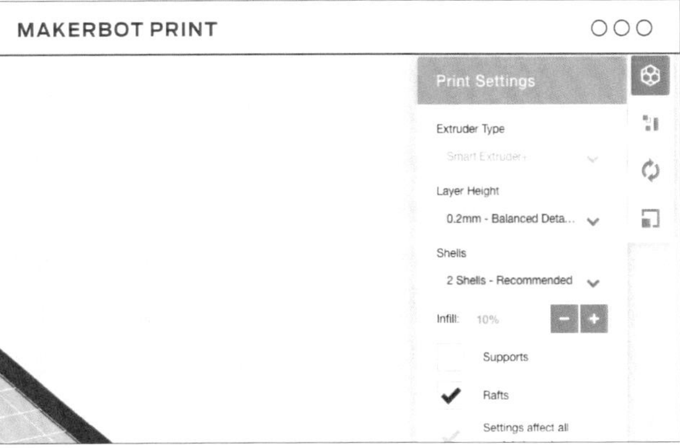

B. Prepare and print your mold box using standard print settings.

STEP 03: SUSPEND MASTER IN MOLD BOX AND PREPARE

You can suspend your master using popsicle sticks, skewers, or 3D printed rods glued to the surface of the master in an inconspicuous place. The holes left in their place after your mold has cured will aid in resin flow through the mold.

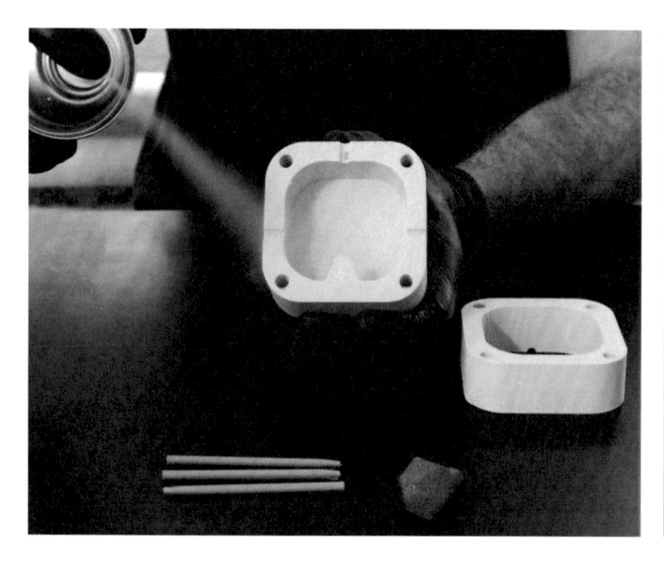

A. Spray your mold box, master, vents, and keys with mold release.

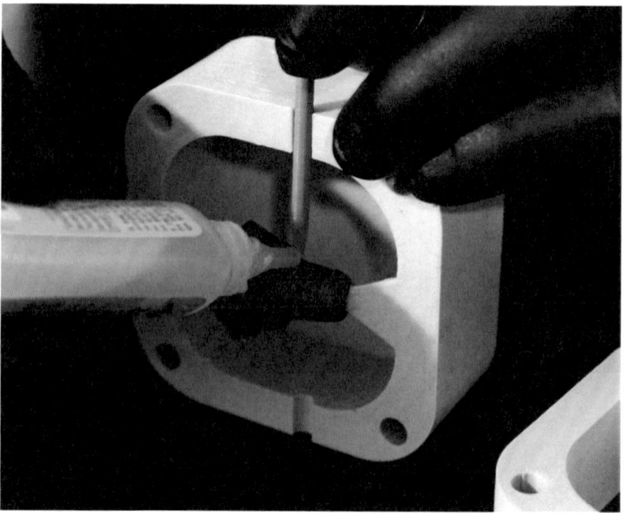

B. Choose points across model to glue vents.

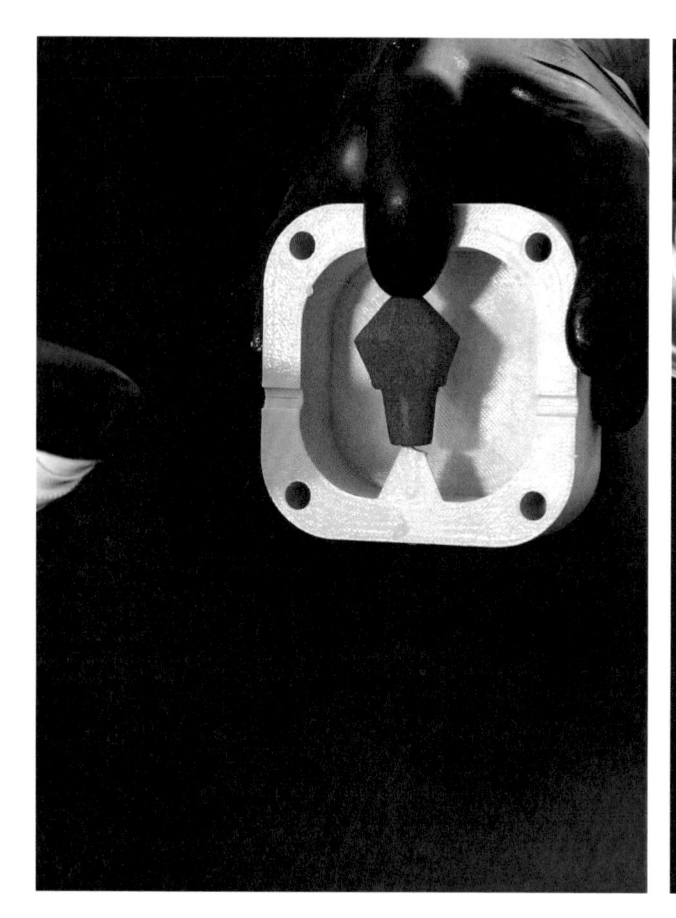

C. Glue the master to the mold box.

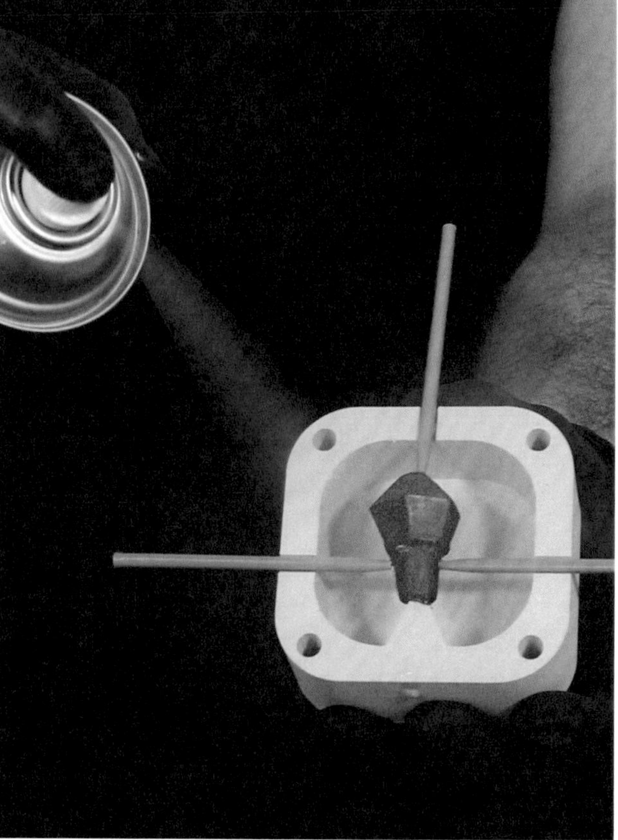

D. Spray again with mold release for good measure.

STEP 04: OPEN BOTTLES OF SILICONE (PARTS A AND B) AND STIR THOROUGHLY

We calculated our mold volume by filling our mold box with water and pouring the water into a measuring cup to find exact volume.

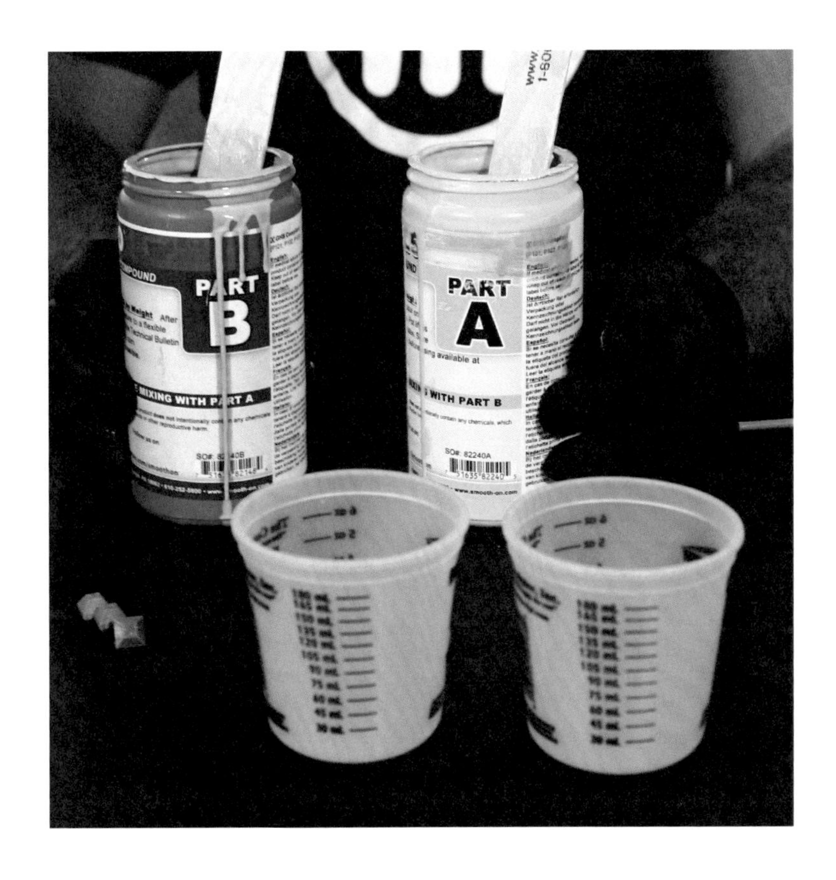

STEP 05: MEASURE

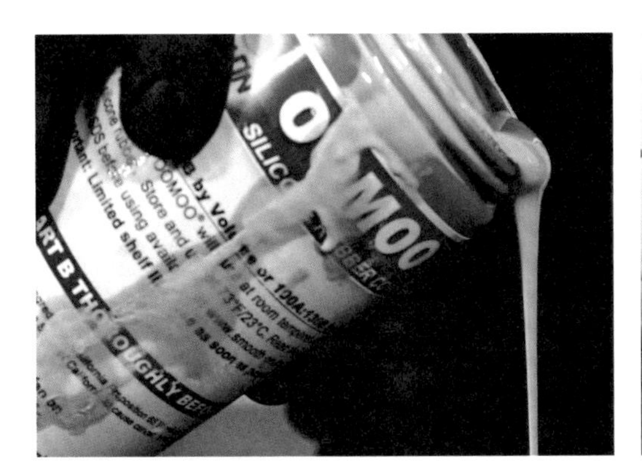

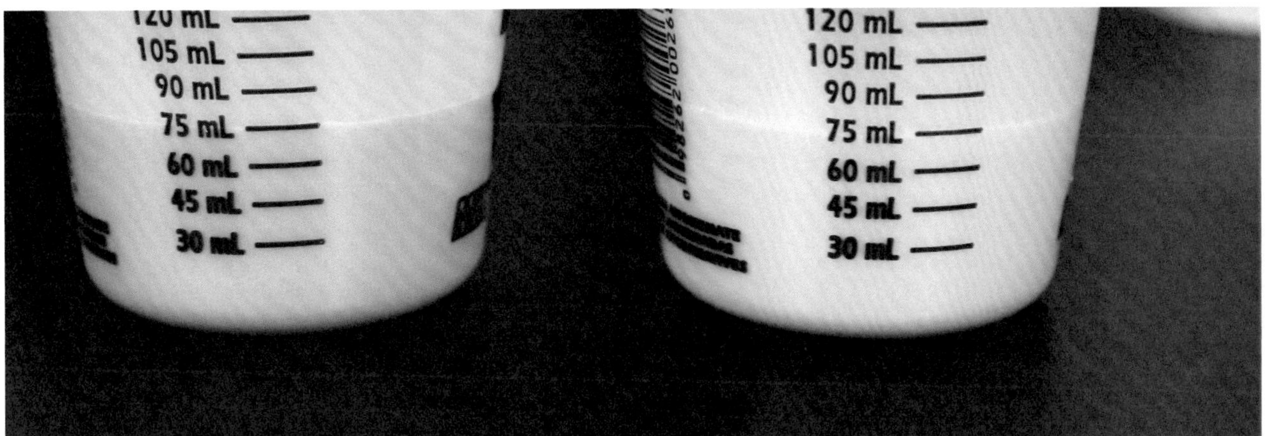

A. Determine the volume of silicone needed to fill your intended mold.

B. Measure and pour the desired amount of silicone and hardener separately in two measuring cups.

STEP 06: MIX SILICONE

Combine the two parts into one mixing cup and stir slowly with a mixing tool. Be cautious not to stir in air bubbles. Scrape the sides of the cup to mix in all material. Once your parts are thoroughly mixed the curing process will begin.

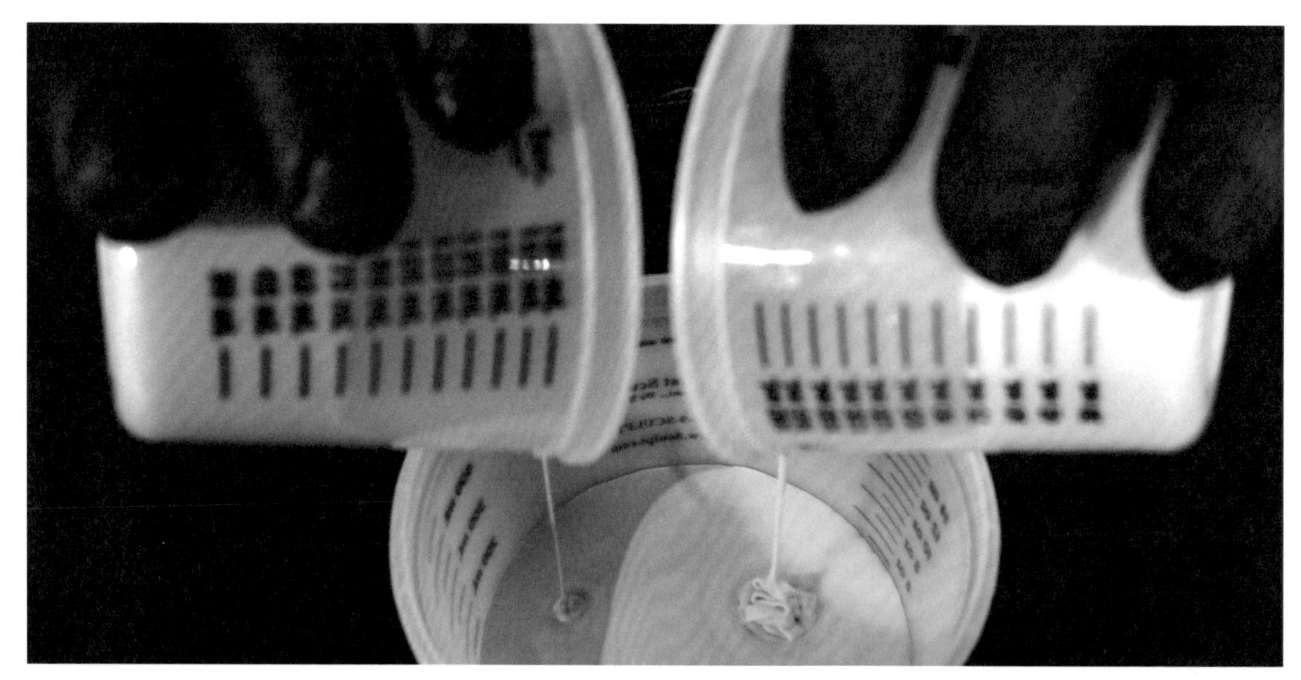

TIP: Read instructions on your silicone for "pot-life." This is how much time you have to work with the silicone before it cures.

STEP 07: POUR MOLD (PART ONE)

Pour silicone into the first half of your mold box. Pour slowly into one corner of the mold box and allow the silicone to run to other parts of the mold box as it fills. Stop when the silicone reaches the top of the box.

After you've poured the silicone, place small "keys" into the uncured silicone. These will create negative spaces and allow the mold halves to fit together once poured. We will remove them before pouring the second half of our mold.

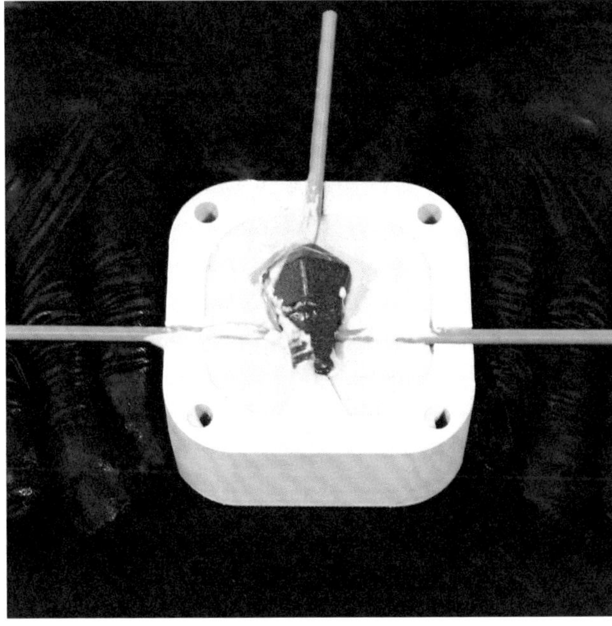

STEP 08: LET CURE

Depending on what type of silicone you are using it can take anywhere from 75 minutes to several hours to cure. Temperature and humidity will affect curing times, so we recommend this process be done in a room temperature environment.

STEP 09: ATTACH AND PREPARE MOLD (PART TWO)

Repeat the process to form the second half of the mold.

A. Remove the keys you inserted in the previous step.

B. Attach and secure the second half of your mold box.

C. Spray with mold release.

STEP 10: POUR INTO MOLD PART TWO

A. Repeat steps 5-8 to create the second half of your mold using the methods mentioned above.

STEP 11: LET CURE

STEP 12: BREAK DOWN MOLD BOX

A. Remove hardware.

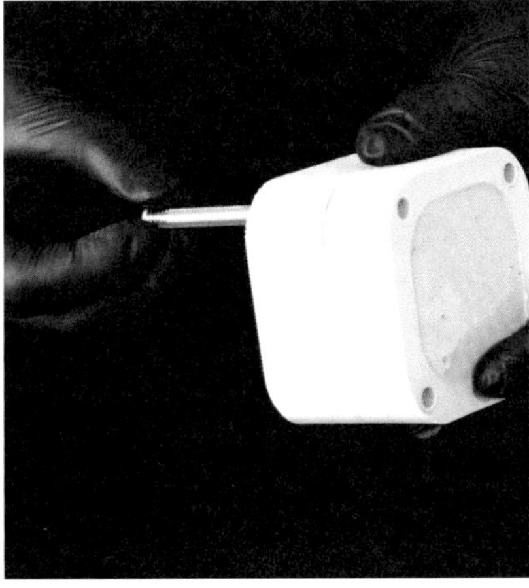

B. Remove the mold from the mold box.

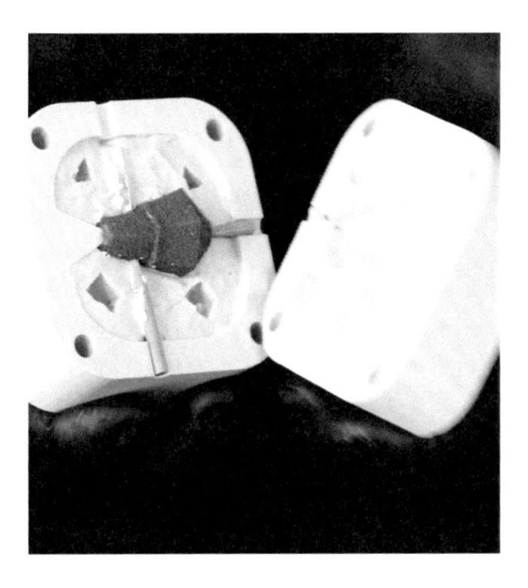

C. Remove the master and vents.

STEP 13: REASSEMBLE AND PREPARE MOLD

A. Ensure that all parts of your mold are correctly aligned, and plug any holes created by vents.

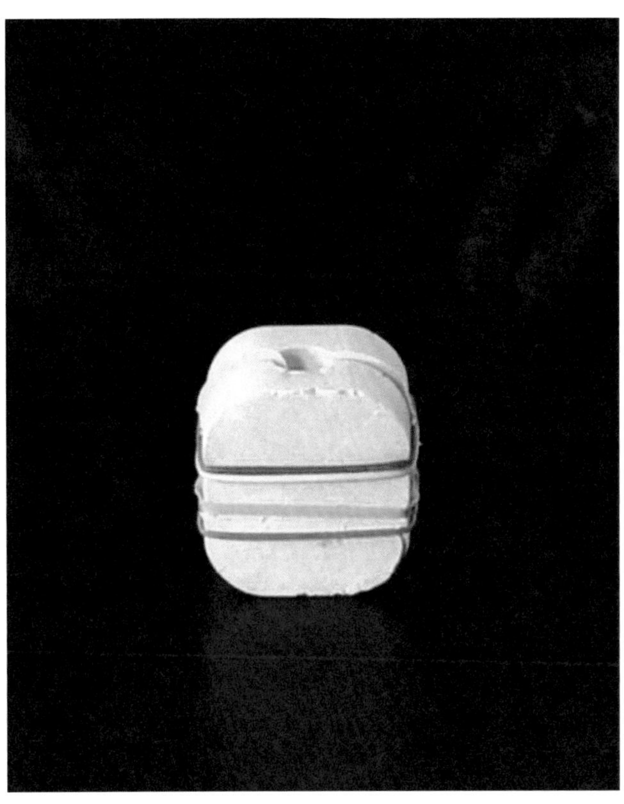

B. Secure the mold pieces using rubber bands, straps, or tape.

TIP: If your vents leave holes in areas where resin can spill out during the pour, they will need to be plugged.

STEP 14: MEASURE RESIN AND ADD DYE

Once you have mixed your resin, you will have only a few short minutes to pour your mold following the steps below.

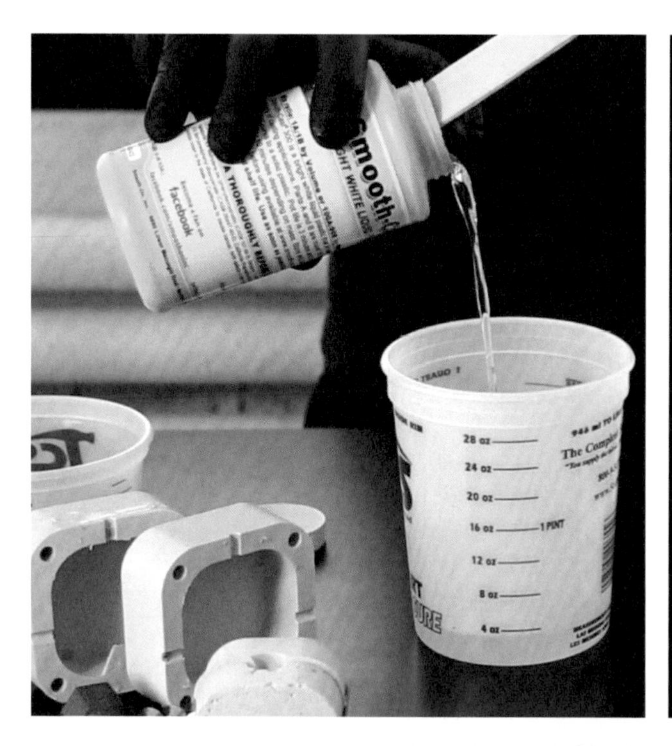

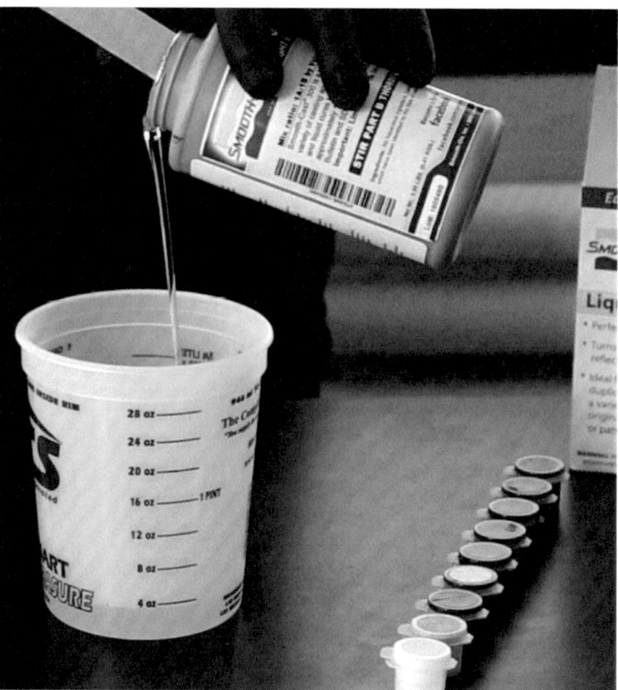

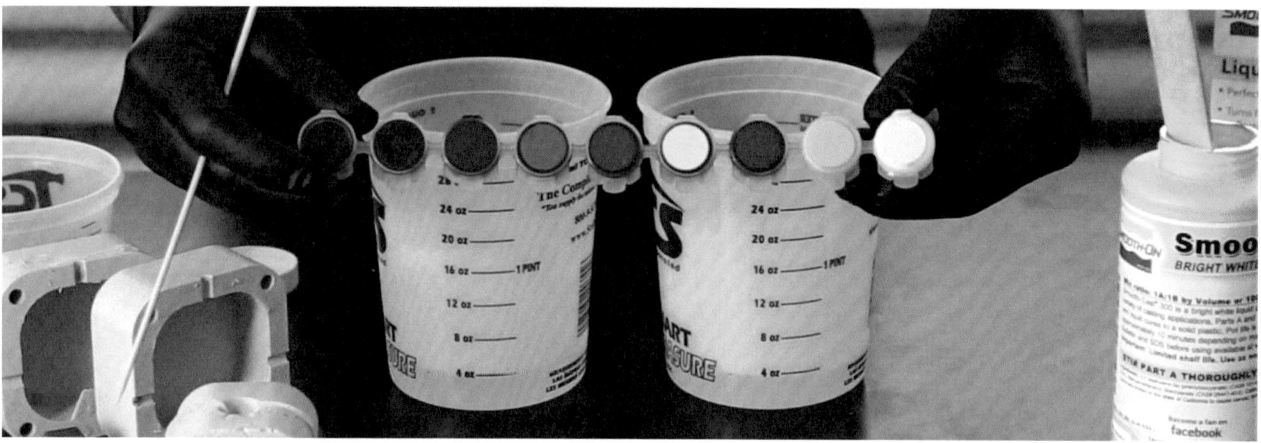

STEP 15: MIX RESIN

A. Combine both parts of the resin mixture and mix thoroughly being sure not to stir in air bubbles.

TIP: Resins typically have a shorter "pot-life" than silicone, meaning they will cure faster.

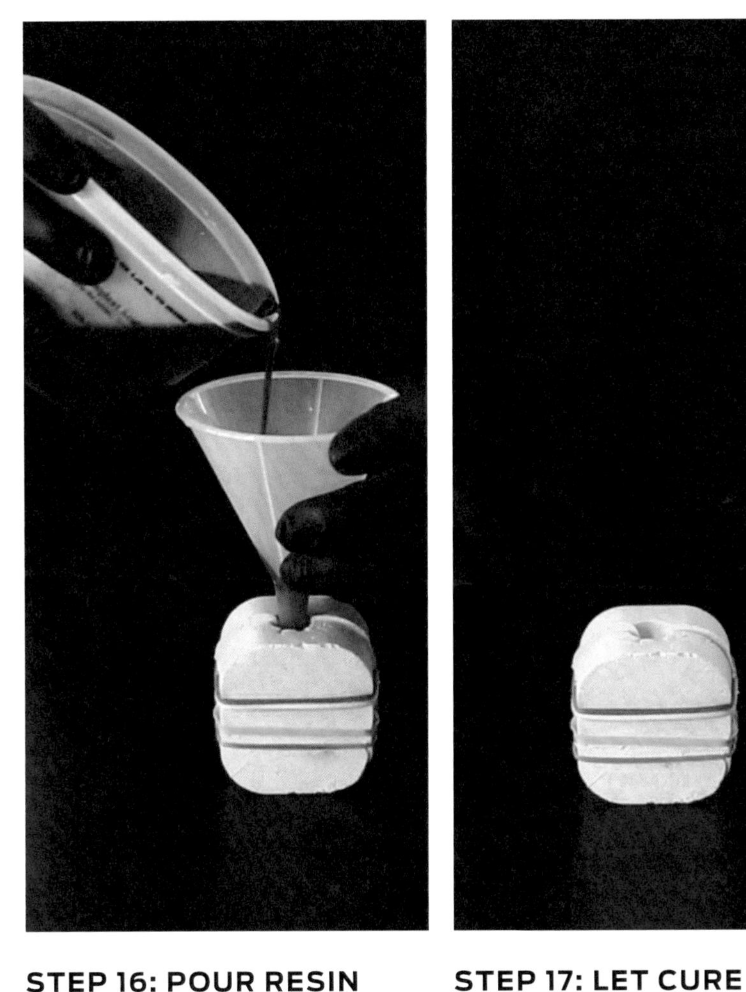

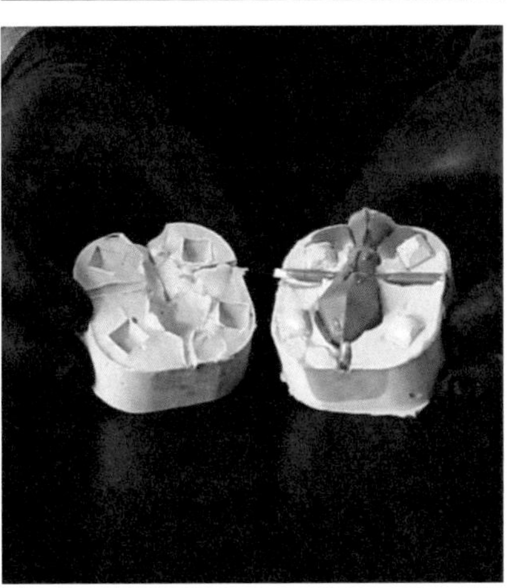

STEP 16: POUR RESIN

Once mixed, pour immediately into the opening of your mold using a funnel. Pour slowly and be careful not to overfill and spill resin.

STEP 17: LET CURE

Once poured, allow the resin to cure for the appropriate amount of time.

STEP 18: OPEN & REMOVE

Once your resin has cured, open the mold and remove your cast part. Any resin that escaped through seams or voids and cured is called "flash". Flash will need to be removed from the part.

STEP 19: FINAL PART

Using this process, you can see that we were able to recreate the perfume bottle cap in several different colors and opacities.

POST-PROCESSING

SILICONE MOLDING WITH 3D PRINTED MOLDS

Silicone molding is a powerful production method when combined with 3D printing. In this How To, we will show you some of the best practices associated with 3D printing molds to pour into. To demonstrate this process, we will create a bicycle handlebar grip out of flexible silicone using a 3D printed mold.

VIDEO TUTORIAL:
makerbot.com/professionals/post-processing/silicone-molding-2

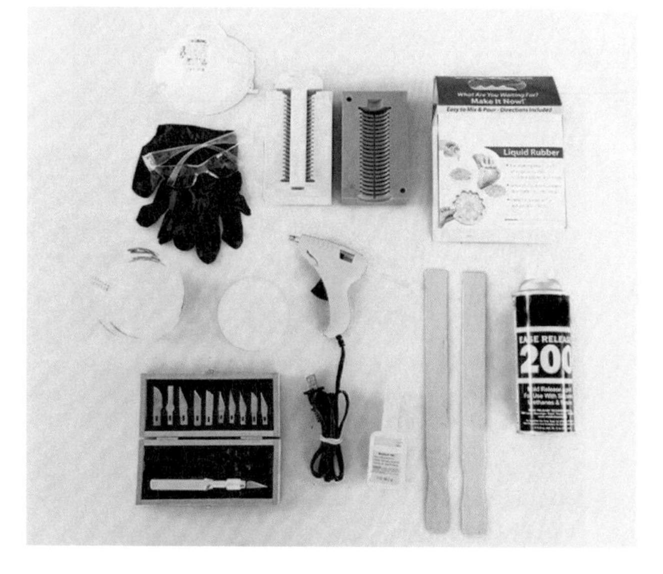

Be sure to consult the safety precautions listed by the manufacturer on any materials and equipment used in this tutorial before starting.

MATERIALS

- 3D printed mold
- Mold release spray
- Measuring/mixing cups
- Mixing sticks
- Hobby knife
- Silicone
- Gloves, eye protection and respiratory mask

STEP 01: PRINT MOLD

BIKE HANDLE BAR GRIP MOLD: *thingiverse.com/thing:103723*

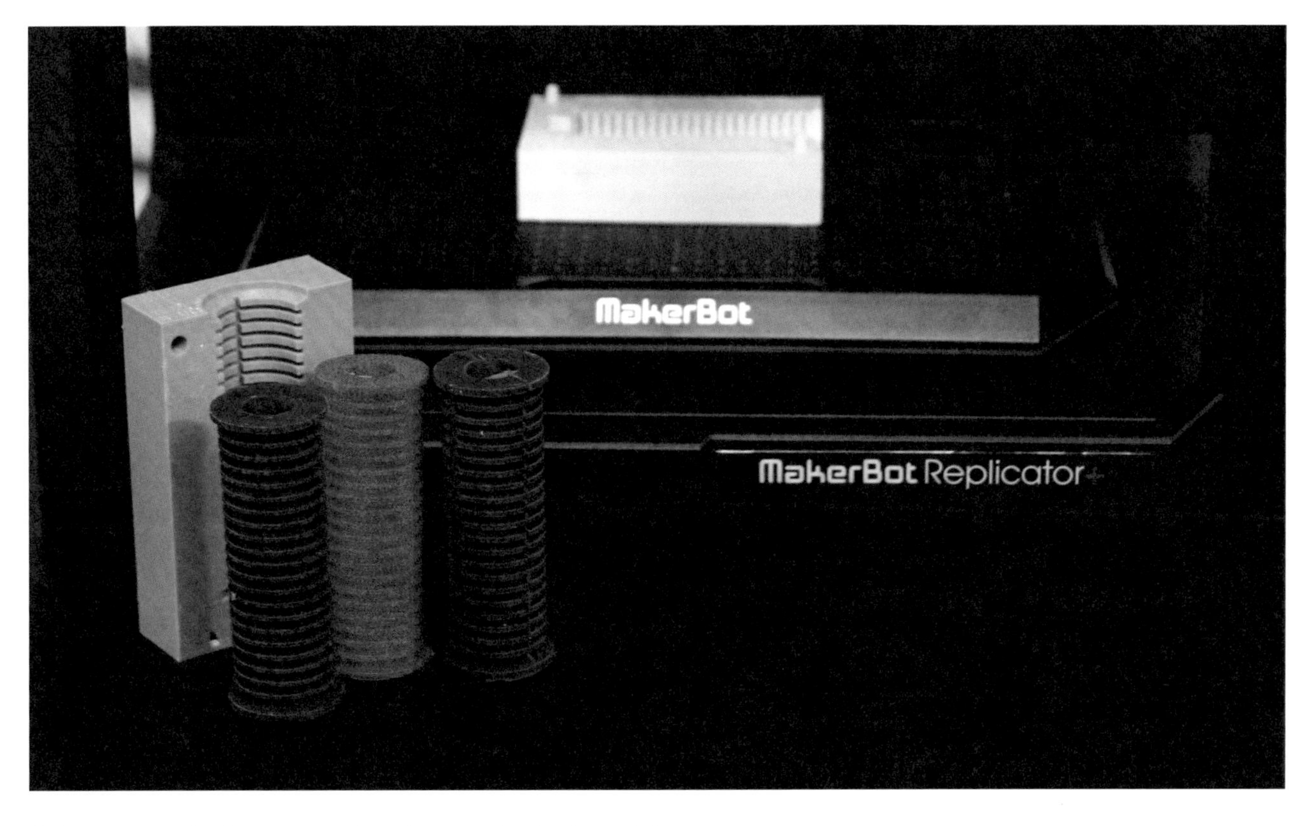

A. Prepare and print your parts using standard print settings.

STEP 02: PREPARE YOUR MOLD

A. Spray all mold surfaces with mold release and fasten mold pieces together with rubber bands.

STEP 03: OPEN, MEASURE, MIX SILICONE AND ADD DYE

Once you've thoroughly mixed your silicone, you can pour it into your printed mold. Because we used a slow curing silicone, we let this mold sit overnight to cure.

A. Measure part A

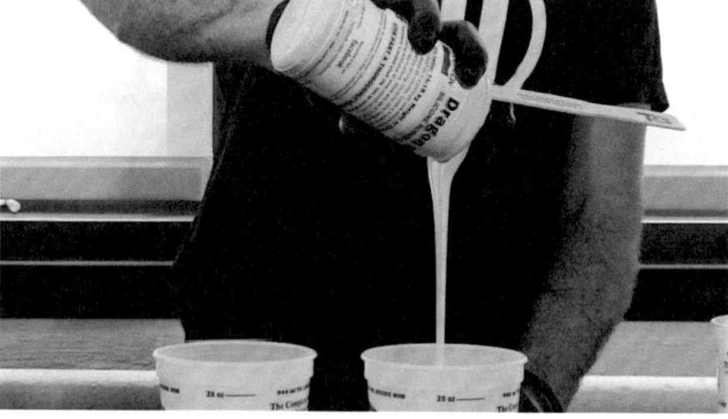

B. Measure part B

C. Add dye to either part A or B (refer to silicone/dye instructions).

D. Stir in dye

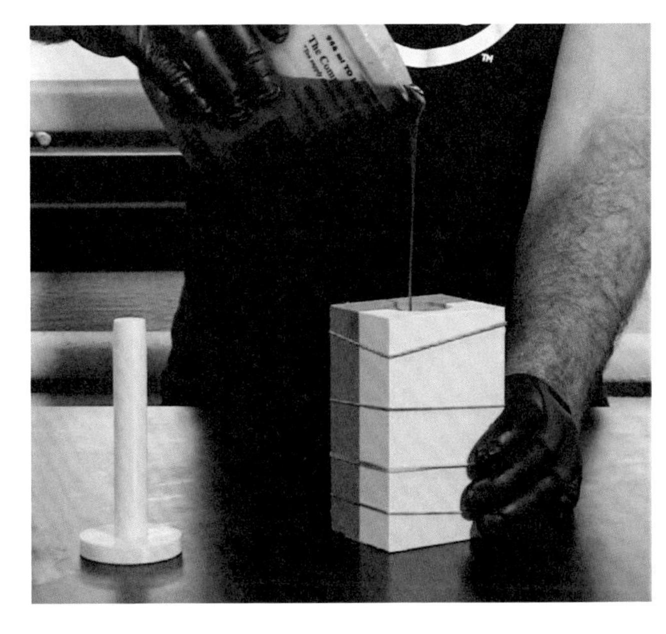

E. Pour mixture into mold.

F. Insert third piece after pouring.

STEP 04: FINAL PART

Using this process, we created a flexible bicycle handlebar grip in a rubber-like material.

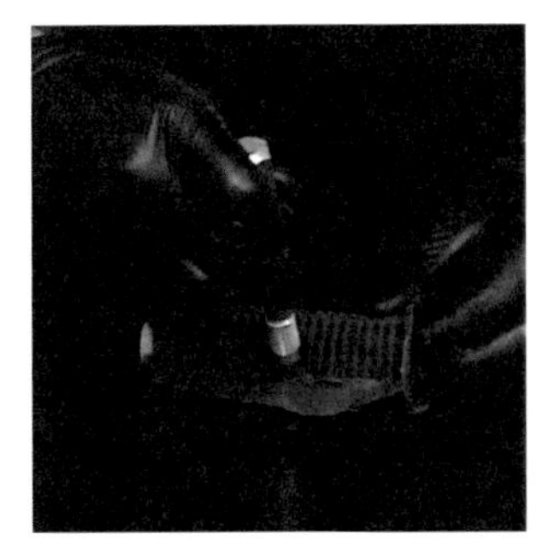

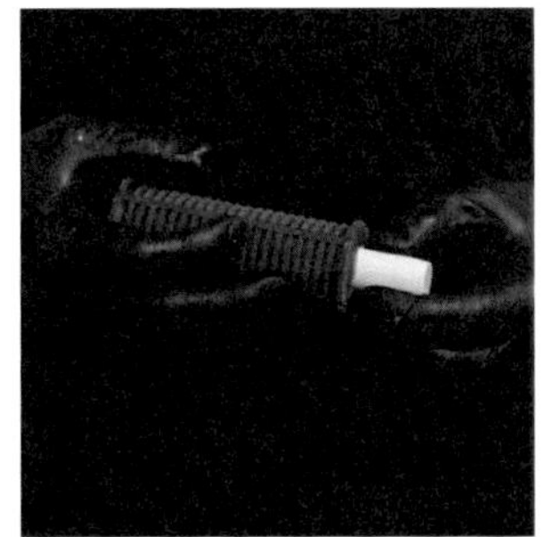

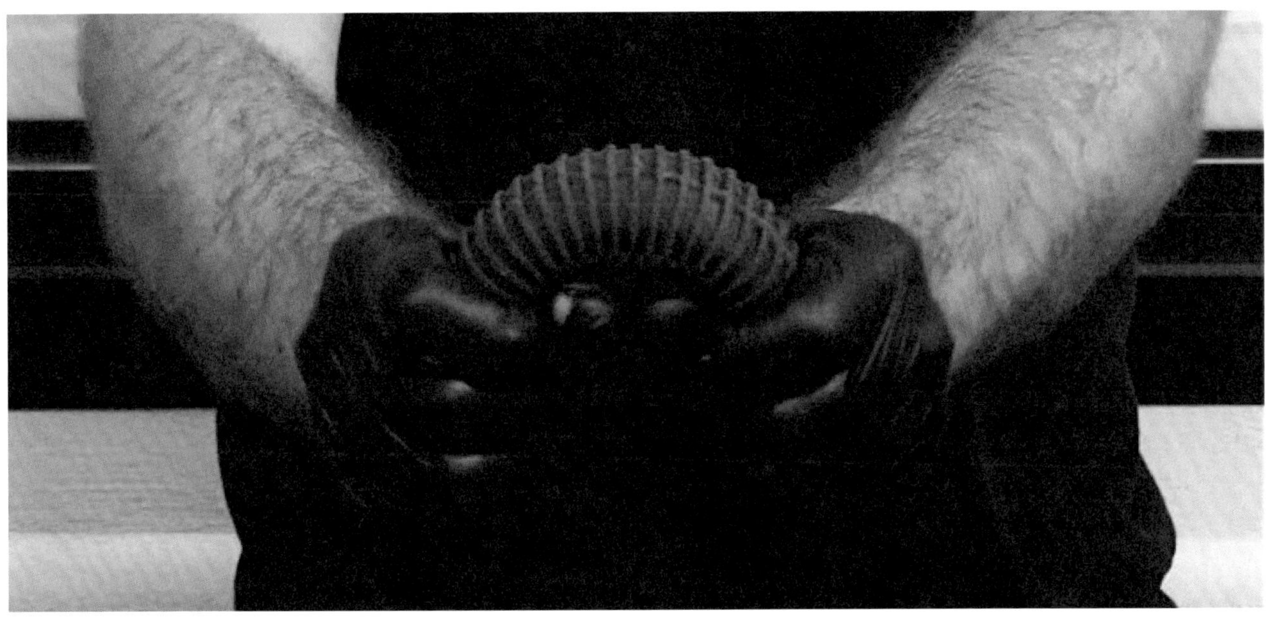

CONGRATULATIONS YOU DID IT!

BY NOW YOU SHOULD KNOW THE FOLLOWING:

Teachers:

› How 3D printers work, hardware and software

› How to setup and maintain your MakerBot 3D printer

› How to integrate 3D printing into a range of class topics

› The different types of 3D design software

› How to create objects in 3D design software

Students:

› How 3D printers work, hardware and software

› The different types of 3D design software

› How to create objects in 3D design software

› The basics of the design process

› Real-world applications of 3D printing and design

BUILDING YOUR OWN LESSON PLANS

Now that you've seen a wide variety of 3D modeling tools, projects, and 3D printing activities, you're perfectly equipped to begin integrating 3D printing into your existing curriculum and begin designing new lessons around what you've learned. Feel free to start slow; look for different touch points in your lessons where 3D printing can easily complement or supplement existing tasks and objects. Even at the most surface level, exposure to 3D printing improves classroom engagement and teaches basic technology concepts. Be sure to check Thingiverse Education often for inspiration and to discover newly uploaded lesson plans.

If you're ready to dive in and create your own lesson plans, focus on the unique things a 3D printer can bring to your classroom.

Custom things - do you have a need for several individually designed things?

Design iteration - is there a place in your lesson for creating different versions of something?

Complex geometries - would using exotic shapes or assemblies be valuable?

As you work on your own ideas for 3D printing projects, share them with the Thingiverse Education community. Your work can be a valuable starting point for other 3D printing beginners who are taking their first steps into a larger world.

KEEP LEARNING KEEP PRINTING

The more you interact with 3D printers and modeling programs, the more amazing applications you'll discover. There's no limit to the age groups, subjects, digital or real-world challenges that can benefit from your growing expertise in the technology.

As you begin to think in 3D, keep in mind that mistakes are an essential part of the process. Each point of failure offers a critical learning opportunity while bringing you one step closer to making something great.

Thanks for reading! We deeply appreciate your attention and interest and will continue to support you in your 3D printing journey.

–The MakerBot Education Team
guidebook@makerbot.com

CONTACT US

Support:

If you have any problems with your printer or the quality of your prints, please visit *makerbot.com/support* to learn more about troubleshooting or file a ticket with our support experts.

MakerBot Educators:

If you have a MakerBot 3D printer and are interested in joining our amazing community of educators, please visit *makerbot.com/educators* and email **education@makerbot.com** with any questions.

Sales:

Have a question about purchasing? Contact **sales@makerbot.com**

Marketing:

Want to partner with MakerBot or share what you've made? Email **pr@makerbot.com** or message us on social media.

twitter.com/makerbot
facebook.com/makerbot

ACKNOWLEDGEMENTS

MakerBot would like to thank the ever-growing community of 3D printing educators and the students they inspire. This guidebook would not have been possible without the support of the innovative teachers that make up the MakerBot Educators and the hundreds of users that generously contribute projects to Thingiverse Education.

AUTHORS

Mair DeMarco
Sean Dippold
Drew Lentz
Josh Snider

PROJECT AUTHORS

Ryan Erickson
Danielle Evansic
Karie Huttner
Kurt Knueve
Poppy Lyttle
Amy Otis
Nichole Thomas
Jeffry Turnmire
Carlos Varas

DESIGNERS

Janice Mercado
Andrew Askedall
Ernst Bernard
Felipe Castaneda

CONTENT CONTRIBUTORS

Tanya Anderson
Lisa Anthony
August Deshais
Tesin Dosch
Douglas Ferguson
Adam Friedman
Alice Gentili
Paul Gerton
Steve Gong
Dennis Gonzalez
Paul Haberstroh
Megan Hendrickson
Yassin Kargbo JR
Christina Keasler
Rebekah Kotlar
Laura Levin
Stephanie Lin
Maria Marsicano
Brian Mernoff
Dale Nance
Ben Nguyen
Elana Reiser
John Riley
Keven Rinaman
Antoinette Schlobohm
Owen Schoeniger
Jessica Simons
Lisa Standring
Bill Steinbach
Aubree Stephens
Katherine Tansey
Iliana Vazuka
Larry Winterlin

CREATIVE CONTRIBUTORS

Vishnu Anantha
Senecca Bratcher
Kiley Boehler
Noa Dayan
Nate Daubert
Lucas Deichl
Megan Gaglio
Ziggy Leacock
Allison Munson
Ana Maria Swiney

SPECIAL THANKS

Erza Cordavi
Ban Dreyre
Alfredo Ganaden
Katie Martell
Jason Nicholas
Caroline Yesquab

TRADEMARKS

The MakerBot "M" logo, MakerBot, MakerBot Education, MakerBot Mobile, MakerBot Print, MakerBot Tough PLA, Thingiverse, and Thingiverse Education are trademarks or registered trademarks of MakerBot Industries, LLC. All rights reserved.

FDM is a registered trademark of Stratasys, Inc.

Mac and iPad are registered trademarks of Apple Inc.

Autodesk, Autodesk Inventor, Fusion 360, Autodesk 123D, Maya, Meshmixer, Mudbox, Tinkercad, and 3Ds MAX, are trademarks or registered trademarks of Autodesk, Inc.

Blender is a registered trademark of the Blender Foundation.

Solidworks is a registered trademark of Dassault Systèmes SolidWorks Corporation.

Dropbox is a registered trademark of Dropbox, Inc.

Google Classroom, Google Drive, and YouTube are trademarks or registered trademarks of Google Inc.

Cinema 4D is a registered trademark of MAXON Computer GmbH.

Minecraft is a registered trademark of Mojang Synergies AB.

Onshape is a registered trademark of Onshape Inc.

Sculptris and ZBrush are trademarks or registered trademarks of Pixologic, Inc.

Sphero and Sprk+ are registered trademarks of Sphero, Inc.

Morphi is a registered trademark of The Inventery, Inc.

Sketchup is a registered trademark of Trimble Inc.

All other brand names, product names or trademarks belong to their respective holders. All rights reserved.

SCREENSHOTS

Tinkercad™ and Fusion 360™ software screenshots reprinted courtesy of Autodesk, Inc.

Minecraft® software screenshots used with permission from Mojang Synergies AB.

All other screenshots belong to their respective holders.

MakerBot Educators Guidebook is an official product of MakerBot Industries, LLC, and is not authorized, sponsored, associated with, or otherwise associated with any of the other parties listed above in this Legal section or otherwise mentioned in this book.